Progress in Scientific Computing
Vol. 1

Edited by
S. Abarbanel
R. Glowinski
G. Golub
H.-O. Kreiss

Springer Science+Business Media, LLC

M. Bernadou
J. M. Boisserie

The Finite Element Method in Thin Shell Theory:
Application to Arch Dam Simulations

1982 Springer Science+Business Media, LLC

Authors:

Michel Bernadou
INRIA
Domaine de Voluceau-Rocquencourt
B.P. 105
F-78153 Le Chesnay Cedex
FRANCE

Jean-Marie Boisserie
E.D.F.-D.E.R.
6, Quai Watier
F-78400 Chatou
FRANCE

CIP-Kurztitelaufnahme der Deutschen Bibliothek

Bernadou, Michel:
The finite element method in thin shell theory :
application to arch dam stimulations /
M. Bernadou ; J.M. Boisserie.
Boston ; Basel ; Stuttgart : Birkhäuser, 1982.
(Progress in scientific computing ; Vol.1)
ISBN 978-0-8176-3070-6 ISBN 978-1-4684-9143-2 (eBook)
DOI 10.1007/978-1-4684-9143-2

NE: Boisserie, Jean-Marie.; GT

Library of Congress Cataloging in Publication Data

Bernadou, M. (Michel), 1943-
The Finite element method in thin shell theory.
(Progress in scientific computing ; v.)
Bibliography: p.
Includes index.
1. Finite element method. 2. Shells (Engineering)
3. Arch dams--Mathematical models. I. Boisserie, J.-M.
(Jean-Marie), 1932- . II. Title. III. Series.
TA347.F5B47 627'.82 82-4293
AACR2

Originally published by Birkhäuser Boston in 1982.

TABLE OF CONTENTS

PREFACE

This Monograph has two objectives : to analyze a *finite element method* useful for solving a large class of *thin shell problems*, and to show in practice how to use this method to simulate an *arch dam problem*.

The first objective is developed in Part I. We record the definition of a general thin shell model corresponding to the W.T. KOITER linear equations and we show the existence and the uniqueness for a solution. By using a *conforming finite element method*, we associate a family of discrete problems to the continuous problem ; prove the convergence of the method ; and obtain error estimates between exact and approximate solutions. We then describe the *implementation* of some specific conforming methods.

The second objective is developed in Part 2. It consists of applying these finite element methods in the case of a representative practical situation that is an *arch dam problem*. This kind of problem is still of great interest, since hydroelectric plants permit the rapid increase of electricity production during the day hours of heavy consumption. This regulation requires construction of new hydroelectric plants on suitable sites, as well as permanent control of existing dams that may be enlightened by numerical stress analysis.

ACKNOWLEDGEMENTS

The authors take this opportunity to express their gratitude to Professors J.L. LIONS, P.G. CIARLET and R. GLOWINSKI for providing all facilities for the development of their researches in an excellent scientific atmosphere at the "Institut National de Recherche en Informatique et en Automatique" (INRIA). They are also indebted to Pr. J.T. ODEN, who has been kind enough to read the manuscript in its entirety and to suggest various improvements ; and to M. LEROY for supplying specifications of GRAND'MAISON arch dam project and for his constant interest.

Many thanks are due to "Electricité de France, Direction des Etudes et Recherches", for constant support and computing facilities.

The authors gratefully appreciate the excellent typing of Mrs. DESNOUS, as well as the kind assistance of the staff of BIRKHAÜSER Boston, in particular that of Ms. K. STEINBERG.

PART I : NUMERICAL ANALYSIS OF A LINEAR THIN SHELL MODEL

Introduction :

A *shell* is a three-dimensional continuous medium for which one dimension, the *thickness*, is "small" with respect to the two others. Under the action of sufficiently small loads, the shell, initially unconstrained, is deformed following the usual laws of the *three-dimensional elasticity*. The basic idea of a *first family of shell theories* is to take into account the particular geometry of such a medium and, by "*integration through the thickness*" to obtain a two-dimensional model, formulated in terms of the *middle surface of the shell*, which represents a "good" approximation of the three-dimensional model. The pioneers of this kind of derivation are KIRCHHOFF [1876] and LOVE [1934]. Their theories were developed and improved by numerous authors, especially by KOITER [1966, 1970] and KOITER and SIMMONDS [1973]. An impressive attempt to derive a unifying approach to the variety of *thick and thin elastic shell theories* and problems has been done recently by RUTTEN [1973]. The mathematical analysis of such derivation methods is now in progress, especially with the works of CIARLET and DESTUYNDER [1979] for plate problems and DESTUYNDER [1980] for shell problems.

A *second* very well-known *family of shell theories* is based on the *COSSERAT* [1909] *surface theory* ; it has been developed by NAGHDI [1963, 1972] among others. Though the basic ideas of these two families are different, their numerical analysis is very similar. Thus, using the methods outlined in Chapter 1, COUTRIS [1976, 1978] completed a numerical analysis of certain shell problems based on NAGHDI's models.

Throughout this book, we will use the KOITER model, which originates from the displacement formulation of three-dimensional elasticity. By means of suitable assumptions about the types of loads applied to the shell and on stress distribution, KOITER has obtained a two-dimensional formulation in terms of geometrical properties of the middle surface of the shell, for which the unknown is *the displacement field of the particles comprising the middle surface*. From the knowledge of this displacement field, one deduces the displacement field and the stress field for any particle of the shell.

In the following, we will discuss the variational formulation of KOITER's model. By defining the middle surface of the shell as the image of a bounded open set of the plane, *the reference domain*, by a regular mapping $\vec{\phi}$, *the problem is henceforward set on the plane reference domain*. Practically, this property is very interesting : *the implementation of a finite element method on a plane domain is more simple than the implementation of finite element methods on any bounded open set of the usual euclidean space*. This fact explains why engineers such as ARGYRIS, HAASE, and MALEJANNAKIS [1973], DUPUIS [1971], DUPUIS and GOËL [1970a, 1970b] and many others, have employed shell theories to analyze practical structural problems. However, a significantly more complex system of governing equations is obtained when reducing to the two-dimensional model ; in addition to the thickness, YOUNG's modulus and POISSON's coefficient (constant or variable), for isotropic shells, such a reduction leads to complicated variable coefficients. The complexity of these equations is responsible for much of the difficulties encountered in the approximation of the solutions of shell problems.

Chapter 1 contains a description of the variational formulation of KOITER's model as well as convenient expressions for the bilinear and linear forms with respect to the approximation. We conclude this chapter by giving broad outlines of the existence and uniqueness theorem for a solution to the problem.

In Chapter 2, we show how to approximate the solution of KOITER's equations by using *conforming finite element methods* and *numerical integration techniques*. Moreover, we prove *convergence* and obtain *estimates of the error* between the exact and approximated solutions. We emphasize that, on the one hand, our study is applicable to *general linear shell equations* using *any system of curvilinear coordinates*, and that, on the other hand, our study provides *criteria* for the choice of suitable numerical integration schemes. This last point is apparently new : up to now, the choice of numerical integration schemes seemed to be based on *empirical* considerations.

The aim of Chapter 3 is to give a detailed description of how to implement the finite element methods considered in Chapter 2. In particular, we emphasize the *modular character* of such an implementation (a module is a set of subroutines) and explicitly describe the following modules: (i) the *interpolation* module, (ii) the *energy functional* module and (iii) the *potential energy of exterior loads* (i.e., second member) module.

The contents of these three chapters are *general* and can be used in many practical problems. In order to illustrate this general matter with a *representative practical situation*, in Part II we consider an application to *arch dam simulations*.

Some notations

Throughout this book, we shall frequently make use of the properties of the *SOBOLEV spaces*. Let Ω be an open bounded subset in a plane $\mathcal{E}^2$. Then, we set

$$\left.\begin{aligned} & W^{m,p}(\Omega) = \{v \in L^p(\Omega) : D^\alpha v \in L^p(\Omega) \quad \text{for } |\alpha| \le m\} \\ & m \ge 1 \quad \text{integer}, \quad 1 \le p < \infty , \end{aligned}\right\}$$

with the usual extension to the case $p = +\infty$. When equipped with the norm

$$\|v\|_{m,p,\Omega} = \left(\sum_{|\alpha|\le m} \int_\Omega |D^\alpha v|^p \, d\xi^1 \, d\xi^2\right)^{1/p}$$

$W^{m,p}(\Omega)$ is a BANACH space. Here ξ^1, ξ^2 denote a system of orthonormal coordinates of the $\mathcal{E}^2$-plane. The corresponding semi-norm is

$$|v|_{m,p,\Omega} = \left(\sum_{|\alpha| = m} \int_\Omega |D^\alpha v|^p \, d\xi^1 \, d\xi^2\right)^{1/p} .$$

In the following, for the case $p = 2$, we shall write $\|v\|_{m,\Omega}$ and $|v|_{m,\Omega}$ instead of $\|v\|_{m,2,\Omega}$ and $|v|_{m,2,\Omega}$. In particular, the space $W^{m,2}(\Omega) = H^m(\Omega)$ is a HILBERT space when endowed with the scalar product

$$((u,v))_{m,\Omega} = \sum_{|\alpha|\leq m} \int_\Omega D^\alpha u \, D^\alpha v \, d\xi^1 \, d\xi^2.$$

In view of section 2.4, we record here some basic properties of the SOBOLEV spaces. The notation $X \hookrightarrow Y$ indicates that the normed linear space X is contained in the normed linear space Y with a continuous injection. By *the SOBOLEV's imbedding theorems*, the following inclusions hold, for all integers $m \geq 0$ and all $1 \leq p \leq \infty$:

$$\left.\begin{array}{l} W^{m,p}(\Omega) \hookrightarrow L^{p^*}(\Omega) \text{ with } \dfrac{1}{p^*} = \dfrac{1}{p} - \dfrac{m}{n}, \quad \text{if } m < \dfrac{n}{p} \\[2ex] W^{m,p}(\Omega) \hookrightarrow L^q(\Omega) \text{ for all } q \in [1,\infty[\quad, \text{ if } m = \dfrac{n}{p} \quad, \\[2ex] W^{m,p}(\Omega) \hookrightarrow \mathcal{C}(\bar{\Omega}) \quad \text{if } \dfrac{n}{p} < m \quad . \end{array}\right\}$$

For more details on SOBOLEV spaces, we refer to ADAMS [1975], LIONS and MAGENES [1968], NEČAS [1967] and ODEN and REDDY [1976].

CHAPTER 1

THE CONTINUOUS PROBLEM

Orientation :

W.T. KOITER's linear theory of thin elastic shells makes use of intrinsic geometrical properties of middle surface of the undeformed shell. Our first task is to define this middle surface and to record relevant results on differential geometry needed in the following. Next, from the knowledge of the geometry of the middle surface, we give the definition of the undeformed shell. Thus, we are able to give a variational formulation of KOITER's model.

With respect to approximations and to applications, we introduce some other convenient expressions pertinent to the above formulation. We conclude with broad outlines of theorems on the existence and uniqueness of a solution to the problem ; the reader primarily interested in applications can skip this last section and proceed directly to the next chapter.

1.1. Definition of the middle surface :

Let $\mathcal{E}^3$ be the usual euclidean space referred to an *orthonormal fixed system* $(0, \vec{e}_1, \vec{e}_2, \vec{e}_3)$, and let Ω be a bounded open subset in a plane $\mathcal{E}^2$, with a boundary Γ. Then, the *middle surface* $\bar{S}$ of the shell is the image in $\mathcal{E}^3$ of the set $\bar{\Omega}$ by mapping $\vec{\phi}$:

$$\vec{\phi} \ : (\xi^1,\xi^2) \in \bar{\Omega} \subset \mathcal{E}^2 \rightarrow \vec{\phi}\ (\xi^1,\xi^2) \in \mathcal{E}^3. \tag{1.1.1}$$

We denote $\partial S = \vec{\phi}(\Gamma)$, hence $\bar{S} = S \cup \partial S$, and we assume $\vec{\phi}$ and Γ sufficiently smooth. Particularly, we assume that all the points of the surface $\bar{S} = \vec{\phi}(\bar{\Omega})$ are *regular* so that the vectors

$$\vec{a}_\alpha = \vec{\phi}_{,\alpha} = \frac{\partial\vec{\phi}}{\partial\xi^\alpha}\ , \quad \alpha = 1,2, \tag{1.1.2}$$

are linearly independent for all points $\xi = (\xi^1,\xi^2) \in \bar{\Omega}$. These two vectors define the *tangent plane* to the surface $\bar{S}$ at the point $\vec{\phi}(\xi)$. Next, we define the *normal vector*

$$\vec{a}_3 = \frac{\vec{a}_1 \times \vec{a}_2}{|\vec{a}_1 \times \vec{a}_2|}\ , \tag{1.1.3}$$

$|\cdot|$ denoting the euclidean norm in $\mathcal{E}^3$ equipped with its usual scalar product $(\vec{a},\vec{b}) \to \vec{a}\cdot\vec{b}$. Then, the point $\vec{\phi}(\xi)$ and the three vectors $\vec{a}_i$ define a *local reference system* for the middle surface (see Figure 1.1.1).

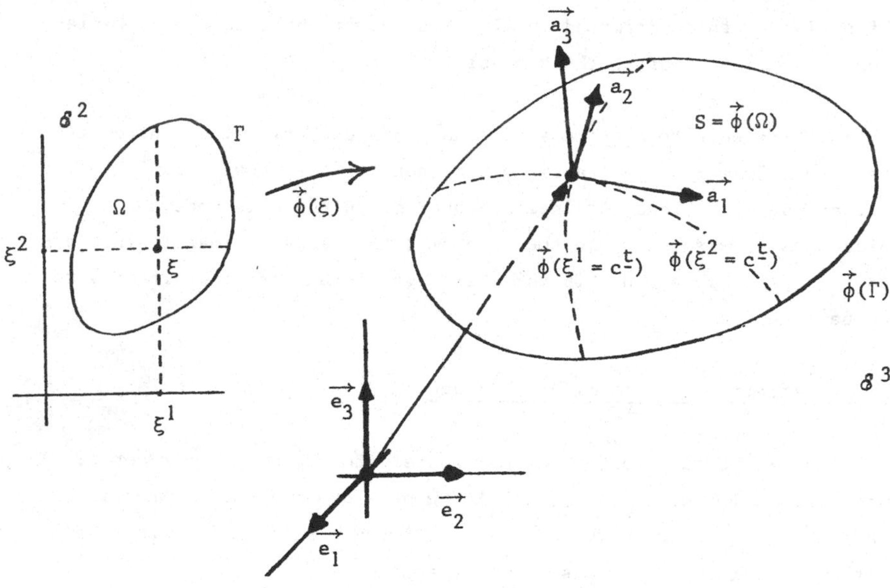

Figure 1.1.1 : Definition of the middle surface $\bar{S}$

In the following, we denote by $a_{\alpha\beta}$, $b_{\alpha\beta}$ the *first and second fundamental forms* of the middle surface S ; that is,

$$a_{\alpha\beta} = a_{\beta\alpha} = \vec{a}_\alpha \cdot \vec{a}_\beta = \vec{\phi}_{,\alpha} \cdot \vec{\phi}_{,\beta} , \tag{1.1.4}$$

$$b_{\alpha\beta} = b_{\beta\alpha} = - \vec{a}_\alpha \cdot \vec{a}_{3,\beta} = \vec{a}_3 \cdot \vec{a}_{\alpha,\beta} = \vec{a}_3 \cdot \vec{a}_{\beta,\alpha} . \tag{1.1.5}$$

As a rule, we shall use Greek letters, $\alpha,\beta,\ldots$, for indices which take their values in the set $\{1,2\}$; Latin letters, $i,j,\ldots$, will be used for indices that take their values in the set $\{1,2,3\}$. In addition, we shall employ the summation convention for a repeated index, occuring once as a subscript and once as a superscript.

To the vectors $\vec{a}_\alpha$, we associate two other vectors $\vec{a}^\beta$ of the tangent plane defined by

$$\vec{a}_\alpha \cdot \vec{a}^\beta = \delta_\alpha^\beta , \quad \delta_\alpha^\beta = \begin{cases} 1 \text{ if } \alpha = \beta, \\ 0 \text{ if } \alpha \neq \beta. \end{cases} \tag{1.1.6}$$

These vectors are linked to the vectors $\vec{a}_\alpha$ by the relations

$$\vec{a}_\alpha = a_{\alpha\beta}\vec{a}^\beta , \quad \vec{a}^\alpha = a^{\alpha\beta}\vec{a}_\beta, \quad a^{\alpha\beta} = \vec{a}^\alpha \cdot \vec{a}^\beta = a^{\beta\alpha} , \tag{1.1.7}$$

where the matrix $(a^{\alpha\beta})$ is the inverse of the matrix $(a_{\alpha\beta})$. This inverse matrix is well defined, since all the points of the middle surface S are assumed to be regular.

For a given tensor, the metric tensors $(a_{\alpha\beta})$ and $(a^{\alpha\beta})$ permit us to derive the different kinds of components. For instance, to the *covariant components* $b_{\alpha\beta}$ of the second fundamental form, we can associate the following *mixed and contravariant components*

$$\left.\begin{aligned} & b_\alpha^{\cdot\beta} = b^\beta_{\cdot\alpha} = b_\alpha^\beta = a^{\beta\lambda} b_{\lambda\alpha} , \\ & b^{\alpha\beta} = a^{\alpha\lambda} a^{\beta\nu} b_{\lambda\nu} , \end{aligned}\right\} \tag{1.1.8}$$

and, conversely,

$$b_{\alpha\beta} = a_{\alpha\lambda} b^{\lambda}_{\beta} = a_{\alpha\lambda} a_{\beta\nu} b^{\lambda\nu} \ . \tag{1.1.9}$$

Since the basis $(\vec{a}_1, \vec{a}_2, \vec{a}_3)$ and $(\vec{a}^1, \vec{a}^2, \vec{a}^3)$ are neither normed nor orthogonal, it is somewhat complicated to compute their derivatives. Thus, it is convenient to introduce the *CHRISTOFFEL's symbols* $\Gamma^{\alpha}_{\beta\gamma}$ defined by

$$\Gamma^{\alpha}_{\beta\gamma} = \Gamma^{\alpha}_{\gamma\beta} = \vec{a}^{\alpha} . \vec{a}_{\gamma,\beta} = \vec{a}^{\alpha} . \vec{a}_{\beta,\gamma} \quad , \tag{1.1.10}$$

as well as the notion of *covariant derivatives* for a surface tensor. For instance, for tensors of order 1 or 2, we have :

$$\left. \begin{aligned} T_{\alpha|\gamma} &= T_{\alpha,\gamma} - \Gamma^{\lambda}_{\alpha\gamma} T_{\lambda} \ , \\ T^{\alpha}|_{\gamma} &= T^{\alpha}_{,\gamma} + \Gamma^{\alpha}_{\lambda\gamma} T^{\lambda} \ , \end{aligned} \right\} \tag{1.1.11}$$

$$\left. \begin{aligned} T_{\alpha\beta|\gamma} &= T_{\alpha\beta,\gamma} - \Gamma^{\lambda}_{\alpha\gamma} T_{\lambda\beta} - \Gamma^{\lambda}_{\beta\gamma} T_{\alpha\lambda} \ , \\ T^{\alpha}_{.\beta|\gamma} &= T^{\alpha}_{.\beta,\gamma} + \Gamma^{\alpha}_{\gamma\lambda} T^{\lambda}_{.\beta} - \Gamma^{\lambda}_{\beta\gamma} T^{\alpha}_{.\lambda} \ , \\ T^{\alpha\beta}|_{\gamma} &= T^{\alpha\beta}_{,\gamma} + \Gamma^{\alpha}_{\lambda\gamma} T^{\lambda\beta} + \Gamma^{\beta}_{\lambda\gamma} T^{\alpha\lambda} \ . \end{aligned} \right\} \tag{1.1.12}$$

Relations (1.1.5) and (1.1.10) involve

$$\left. \begin{aligned} \vec{a}_{\alpha,\beta} &= \Gamma^{\gamma}_{\alpha\beta} \vec{a}_{\gamma} + b_{\alpha\beta} \vec{a}_3 \\ \vec{a}^{\alpha}_{,\beta} &= - \Gamma^{\alpha}_{\beta\lambda} \vec{a}^{\lambda} + b^{\alpha}_{\beta} \vec{a}_3 \end{aligned} \right\} \quad \text{(GAUSS)} \tag{1.1.13}$$

$$\vec{a}_{3,\alpha} = \vec{a}^3_{,\alpha} = - b^{\gamma}_{\alpha} \vec{a}_{\gamma} \quad \text{(WEINGARTEN)} \tag{1.1.14}$$

Expressions of some cross products of basis vectors are also worth noting :

$$\left. \begin{aligned} \vec{a}_{\alpha} \times \vec{a}_{\beta} &= \varepsilon_{\alpha\beta} \vec{a}^3 \ , \\ \vec{a}^{\alpha} \times \vec{a}^{\beta} &= \varepsilon^{\alpha\beta} \vec{a}_3 \ , \\ \vec{a}_3 \times \vec{a}_{\beta} &= \varepsilon_{\beta\lambda} \vec{a}^{\lambda} \ , \\ \vec{a}_3 \times \vec{a}^{\beta} &= \varepsilon^{\beta\lambda} \vec{a}_{\lambda} \ . \end{aligned} \right\} \tag{1.1.15}$$

Here,

$$\varepsilon_{\alpha\beta} = \sqrt{a}\, e_{\alpha\beta} \quad , \qquad \varepsilon^{\alpha\beta} = \frac{1}{\sqrt{a}}\, e^{\alpha\beta} \quad , \tag{1.1.16}$$

$$(e_{\alpha\beta}) = (e^{\alpha\beta}) = \begin{pmatrix} 0 & 1 \\ -1 & 0 \end{pmatrix} \quad , \tag{1.1.17}$$

$$a = a_{11}a_{22} - (a_{12})^2 \neq 0 \quad \text{(regular points).} \tag{1.1.18}$$

The parameter a appears in the expression of the *area element* dS *of the surface*, that is,

$$dS = |\vec{a}_1 \times \vec{a}_2|\, d\xi^1 d\xi^2 = \sqrt{a}\, d\xi^1 d\xi^2 \quad . \tag{1.1.19}$$

1.2. Geometrical definition of the undeformed shell $\mathcal{C}$:

In addition to the two curvilinear coordinates ξ^1, ξ^2 that allow definition of the middle surface, we introduce a third curvilinear coordinate, ξ^3, which is measured along the normal $\vec{a}_3$ to the surface $\bar{S}$ at point $\vec{\phi}(\xi^1,\xi^2)$. This system (ξ^1,ξ^2,ξ^3) of curvilinear coordinates is, at least locally, a system of curvilinear coordinates of $\mathcal{E}^3$, generally called *normal coordinates system*.

The *thickness* e of the shell is defined through a mapping

$$e : (\xi^1,\xi^2) \in \bar{\Omega} \rightarrow \{x \in \mathbb{R} \; ; \; x > 0\} \; . \tag{1.2.1}$$

Then, the *shell* $\mathcal{C}$ is the closed subset of $\mathcal{E}^3$ defined by

$$\begin{aligned} \mathcal{C} = \{M \in \mathcal{E}^3 \; ; \; \vec{OM} = \vec{\phi}(\xi^1,\xi^2) + \xi^3\vec{a}_3 \quad , \\ (\xi^1,\xi^2) \in \bar{\Omega} \, , \; -\frac{1}{2}\, e\,(\xi^1,\xi^2) \leq \xi^3 \leq \frac{1}{2}\, e\,(\xi^1,\xi^2)\} \end{aligned} \tag{1.2.2}$$

The derivatives of the vector $\vec{OM} = \vec{\phi}(\xi^1,\xi^2) + \xi^3\vec{a}_3$ are vectors $\vec{g}_i$ which, by virtue of (1.1.14) satisfy

$$\left.\begin{aligned} \vec{g}_\alpha &= \vec{OM}_{,\alpha} = (\delta^\nu_\alpha - \xi^3 b^\nu_\alpha)\, \vec{a}_\nu \, , \\ \vec{g}_3 &= \vec{OM}_{,3} = \vec{a}_3 \quad . \end{aligned}\right\} \tag{1.2.3}$$

The vectors $\vec{g}_1$ and $\vec{g}_2$ are parallel to the tangent plane to the middle surface at point $\vec{\phi}(\xi^1,\xi^2)$ while vector $\vec{g}_3$ is normal to this plane (see Figure 1.2.1). In BERNADOU and CIARLET [1976, § 2.1] it is shown that $(\vec{g}_1, \vec{g}_2, \vec{g}_3)$ is a *local reference system* at any point of the shell $\mathcal{C}$.

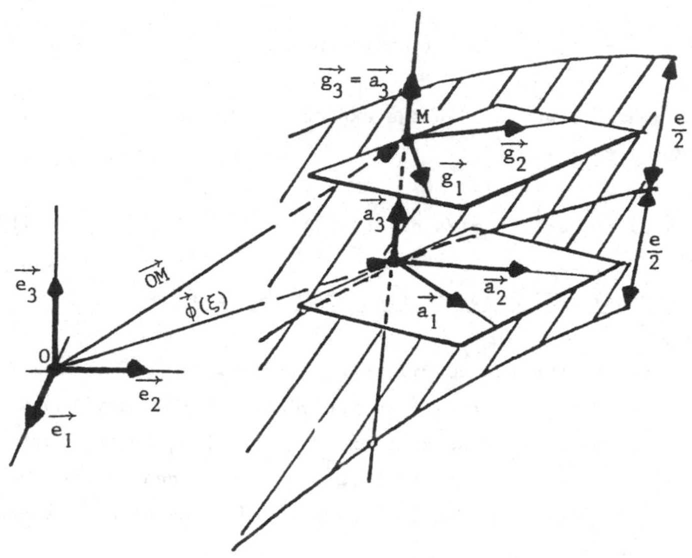

Figure 1.2.1 : Different local basis through the thickness

1.3. The linear model of W.T. KOITER :

From this point onward, the geometry of the shell $\mathcal{C}$, as defined in Section 1.2., is used as a *reference configuration*. We shall study *stationary* problems for which $\mathcal{C}$ is the shell configuration before deformation. We assume that the shell is *clamped* on a part of the boundary and is loaded by some distribution of volume and surface forces. These surface forces act on the upper and lower faces as well as the unclamped part of the boundary. Under the action of these forces, the shell is deformed and takes on a *new configuration* $\mathcal{C}^*$ when it attains static equilibrium. Then, knowing the physical characteristics of the shell material, the initial configuration $\mathcal{C}$, the distribution of the applied forces and the boundary conditions, the problem is to determine the *displacement field* $\vec{U}$ of the particles of $\mathcal{C}$. From the knowledge of the displacements, we are then able to determine the *strains* and the *stresses* at any point of $\mathcal{C}^*$.

The KOITER theory is based on *complementary hypotheses* which permit to derive a satisfactory approximation of the displacement field $\vec{U}$ of the particles of the shell $\mathcal{C}$ from only the knowledge of the *displacement field* $\vec{u}$ *of the particles of its middle surface* S. These hypotheses are listed as follows :

Complementary hypotheses of KOITER [1966, pages 15-16] :

(i) *The normal to the undeformed middle surface, considered as a set of particles of the shell, remains normal to the deformed middle surface ;*

(ii) *During the deformation, the stresses are approximatively plane and parallel to the tangent plane to the middle surface.*

∎

More precisely, let us denote by M the position of some particle of $\mathcal{C}$ and by P the projection of M on the middle surface of $\mathcal{C}$. Let $\bar{M}$ and $\bar{P}$ denote the respective positions of these particles after deformation. With the complementary hypothesis (i), the relation

$$\vec{OM} = \vec{OP} + \xi^3 \vec{a}_3 \qquad (1.3.1)$$

becomes, in the deformed configuration $\mathcal{C}^*$,

$$\vec{O\bar{M}} = \vec{O\bar{P}} + \bar{\xi}^3 \vec{\bar{a}}_3 \quad . \qquad (1.3.2)$$

Indeed, one can check that the vectors

$$\vec{\bar{a}}_\alpha = \vec{O\bar{P}}_{,\alpha} \qquad (1.3.3)$$

are linearly independent ; so, as with we can define the vector

$$\vec{\bar{a}}_3 = \frac{\vec{\bar{a}}_1 \times \vec{\bar{a}}_2}{|\vec{\bar{a}}_1 \times \vec{\bar{a}}_2|} \quad . \qquad (1.3.4)$$

Then, the parameter $\bar{\xi}^3$ defines the coordinate of point $\bar{M}$ measured along $(\bar{P}, \vec{\bar{a}}_3)$. Thus, the deformed configuration $\mathcal{C}^*$ is referred to a *new system of curvilinear coordinates* $(\xi^1,\xi^2,\bar{\xi}^3)$, which differs from the initial system (ξ^1,ξ^2,ξ^3) by the third coordinate.

For a first approximation, it seems reasonable to assume that $\bar{\xi}^3$ is independent of ξ^1 and ξ^2. One can show (see KOITER [1966, page 15] or NAGHDI [1963, § 4]) that hypothesis (ii) permits us to determine $\bar{\xi}^3$. Analogously to (1.2.3), relation (1.3.2) allows us to define at each point $\bar{M}$ of $\mathcal{C}^*$ the vectors

$$\left.\begin{aligned} \vec{\bar{g}}_\alpha &= (\delta^\lambda_\alpha - \bar{b}^\lambda_\alpha \bar{\xi}^3)\, \vec{\bar{a}}_\lambda \,, \\ \vec{\bar{g}}_3 &= \frac{d\bar{\xi}^3}{d\xi^3}\, \vec{\bar{a}}_3 \,, \end{aligned}\right\} \tag{1.3.5}$$

where $\bar{b}^\lambda_\alpha$ denote mixed components of the second fundamental form of the deformed middle surface. In BERNADOU and CIARLET [1976, § 2.3], it is shown that the set $(\vec{\bar{g}}_1, \vec{\bar{g}}_2, \vec{\bar{g}}_3)$ constitutes a *local reference system of* $\mathcal{C}^*$.

Evaluation of the strain tensor of the shell $\mathcal{C}$:

For a three-dimensional continuous medium, the strain tensor is given by

$$\gamma^*_{ij} = \frac{1}{2}\,(\bar{g}_{ij} - g_{ij}) \tag{1.3.6}$$

where $\bar{g}_{ij}$ (respectively, g_{ij}) is the metric tensor of the continuous medium in the deformed (respectively, undeformed) configuration, for a same parameterization (ξ^1,ξ^2,ξ^3). The exponent * which appears in (1.3.6) indicates a difference between tensors in $\mathcal{E}^3$, on the one hand, and tensors defined on the surface, on the other. Relations (1.1.4) (1.2.3) and (1.3.5) lead to the formulas,

$$\left.\begin{aligned} g_{\alpha\beta} &= \vec{g}_\alpha \cdot \vec{g}_\beta = a_{\alpha\beta} - 2b_{\alpha\beta}\xi^3 + b^\lambda_\alpha b_{\lambda\beta}(\xi^3)^2 \,, \\ \bar{g}_{\alpha\beta} &= \vec{\bar{g}}_\alpha \cdot \vec{\bar{g}}_\beta = \bar{a}_{\alpha\beta} - 2\bar{b}_{\alpha\beta}\bar{\xi}^3 + \bar{b}^\lambda_\alpha \bar{b}_{\lambda\beta}(\bar{\xi}^3)^2 \,, \\ g_{\alpha 3} &= g_{3\alpha} = \bar{g}_{\alpha 3} = \bar{g}_{3\alpha} = 0 \,. \end{aligned}\right\} \tag{1.3.7}$$

The assumptions (in this entire theory) of small displacements and small strains, added to that of a thin shell, allow us to neglect the terms in $(\xi^3)^2$ and $(\bar{\xi}^3)^2$ on the one hand, and to substitute ξ^3 for $\bar{\xi}^3$ in $\bar{g}_{\alpha\beta}$ on the other. Then, from (1.3.6), we obtain

$$\left.\begin{aligned} \gamma^*_{\alpha\beta} &= \frac{1}{2}(\bar{a}_{\alpha\beta} - a_{\alpha\beta}) - (\bar{b}_{\alpha\beta} - b_{\alpha\beta})\,\xi^3, \\ \gamma^*_{\alpha 3} &= \gamma^*_{3\alpha} = 0\,. \end{aligned}\right\} \qquad (1.3.8)$$

Moreover, for an elastic homogeneous isotropic material satisfying *HOOKE's law*, we can check that the second complementary hypothesis (ii) of plane stresses implies

$$\gamma^*_{33} = -\frac{\nu}{1-\nu}\,\gamma^{*\alpha}_{\ \alpha} \quad \text{with} \quad \gamma^{*\alpha}_{\ \beta} = g^{\alpha\lambda}\gamma^*_{\beta\lambda}\,, \qquad (1.3.9)$$

where ν is *POISSON's coefficient* of the material. In other words, these relations show that the evaluation of the strain tensor of the shell $\mathcal{C}^*$ depends on the evaluation of the two following surface tensors :

(i) *the strain tensor of the middle surface*

$$\gamma_{\alpha\beta} = \frac{1}{2}(\bar{a}_{\alpha\beta} - a_{\alpha\beta})\,, \qquad (1.3.10)$$

(ii) *the change of curvature tensor of the middle surface*

$$\bar{\rho}_{\alpha\beta} = \bar{b}_{\alpha\beta} - b_{\alpha\beta}\,. \qquad (1.3.11)$$

<u>The displacement field $\vec{U}$ of the particles of the shell</u> $\mathcal{C}$:

With the notations governing relations (1.3.1) (1.3.2), we obtain

$$\vec{U} = \overrightarrow{M\bar{M}} = \overrightarrow{P\bar{P}} + \bar{\xi}^3\vec{\bar{a}}_3 - \xi^3\vec{a}_3\,. \qquad (1.3.12)$$

By

$$\vec{u} = \overrightarrow{P\bar{P}} \qquad (1.3.13)$$

we mean the *displacement field* of the particles of the *middle surface*. For a first approximation, we check that the displacement field $\vec{U}$

depends only on the displacement field $\vec{u}$. We begin by giving the expression of vector $\vec{\bar{a}}_3$ on the basis $(\vec{a}_1, \vec{a}_2, \vec{a}_3)$. From (1.3.13), we have

$$\overrightarrow{O\bar{P}} = \overrightarrow{OP} + \vec{u} = \vec{\phi} + \vec{u} \quad , \tag{1.3.14}$$

so that, with (1.1.2) and (1.3.3),

$$\vec{\bar{a}}_\alpha = \overrightarrow{O\bar{P}}_{,\alpha} = \vec{a}_\alpha + \vec{u}_{,\alpha} \quad . \tag{1.3.15}$$

Relations (1.1.11) (1.1.13) allow us to write the partial derivatives of the vector

$$\vec{u} = u_i \vec{a}^i \quad ,$$

in the form

$$\vec{u}_{,\alpha} = (u_{\lambda|\alpha} - b_{\lambda\alpha} u_3)\, \vec{a}^\lambda + (u_{3|\alpha} + b^\lambda_\alpha u_\lambda)\, \vec{a}^3 \quad , \tag{1.3.16}$$

where we have set

$$u_{3|\alpha} = u_{3,\alpha} \quad . \tag{1.3.17}$$

Then, relations (1.3.15) and (1.3.16) give

$$\vec{\bar{a}}_\alpha = (a_{\lambda\alpha} + u_{\lambda|\alpha} - b_{\lambda\alpha} u_3)\, \vec{a}^\lambda + (u_{3|\alpha} + b^\lambda_\alpha u_\lambda)\, \vec{a}^3 \quad . \tag{1.3.18}$$

Let us substitute these expressions in (1.3.4) and use relations (1.1.15). After linearization, we obtain

$$\vec{\bar{a}}_3 = -\,(u_{3|\alpha} + b^\lambda_\alpha u_\lambda)\, \vec{a}^\alpha + \vec{a}_3 \quad . \tag{1.3.19}$$

Afterwards, for a first approximation, we are allowed to substitute $\bar{\xi}^3$ for ξ^3 into the relation (1.3.12), and so obtain

$$\vec{U} = \vec{u} - \xi^3 (u_{3|\alpha} + b^\lambda_\alpha u_\lambda)\, \vec{a}^\alpha \quad . \tag{1.3.20}$$

Expression of tensors $\gamma_{\alpha\beta}$ and $\bar{\rho}_{\alpha\beta}$ as functions of $\vec{u}$:

After linearization, relations (1.3.10) (1.3.18) give

$$\gamma_{\alpha\beta}(\vec{u}) = \frac{1}{2}(u_{\beta|\alpha} + u_{\alpha|\beta}) - b_{\alpha\beta}u_3 \quad . \tag{1.3.21}$$

Similarly, the differentiation of relation (1.3.18) and the relation (1.3.19) allow us to obtain a linearized form of $\bar{b}_{\alpha\beta} = \vec{a}_3 \cdot \vec{a}_{\alpha,\beta}$ so that, with (1.3.11), we get

$$\bar{\rho}_{\alpha\beta}(\vec{u}) = u_{3|\alpha\beta} - b^{\lambda}_{\alpha}b_{\lambda\beta}u_3 + b^{\lambda}_{\beta}|_{\alpha}u_{\lambda} + b^{\lambda}_{\beta}u_{\lambda|\alpha} + b^{\lambda}_{\alpha}u_{\lambda|\beta} \ , \tag{1.3.22}$$

where

$$u_{3|\alpha\beta} = u_{3,\alpha\beta} - \Gamma^{\lambda}_{\alpha\beta}u_{3,\lambda} \quad . \tag{1.3.23}$$

Shell strain energy associated to a displacement field $\vec{v}$:

The *shell strain energy* associated to a displacement field $\vec{v}$ of the middle surface S can be written as

$$\left.\begin{aligned} a(\vec{v},\vec{v}) = \int_S \frac{Ee}{1-\nu^2}\Big\{ & (1-\nu)\gamma^{\alpha}_{\beta}(\vec{v})\gamma^{\beta}_{\alpha}(\vec{v}) + \nu\gamma^{\alpha}_{\alpha}(\vec{v})\gamma^{\beta}_{\beta}(\vec{v}) + \\ & + \frac{e^2}{12}\Big[(1-\nu)\bar{\rho}^{\alpha}_{\beta}(\vec{v})\bar{\rho}^{\beta}_{\alpha}(\vec{v}) + \nu\bar{\rho}^{\alpha}_{\alpha}(\vec{v})\bar{\rho}^{\beta}_{\beta}(\vec{v})\Big]\Big\} \, dS \quad , \end{aligned}\right\} \tag{1.3.24}$$

where the mixed components $\gamma^{\alpha}_{\beta}(\vec{v})$ and $\bar{\rho}^{\alpha}_{\beta}(\vec{v})$ are given by

$$\gamma^{\alpha}_{\beta}(\vec{v}) = a^{\alpha\lambda}\gamma_{\lambda\beta}(\vec{v}) \ , \ \bar{\rho}^{\alpha}_{\beta}(\vec{v}) = a^{\alpha\lambda}\bar{\rho}_{\lambda\beta}(\vec{v}) \quad , \tag{1.3.25}$$

and where E,ν respectively denote *YOUNG's modulus* and *POISSON's coefficient* of the material. The expression (1.3.24) is an *approximation* of the shell strain energy. Its derivation from the three dimensional elasticity theory is carried out by integration on the thickness of the shell, following the *methods of asymptotic expansions*. For a *formal* utilization of these methods in shell theory, we refer to KOITER [1966], NAGHDI [1972], RUTTEN [1973], and to the bibliography of these works ; for the mathematical justifications, see CIARLET and DESTUYNDER [1979], and DESTUYNDER [1980].

Thus, with the expression (1.3.24), we are able to associate the following symmetric bilinear form, *directly defined on the open plane set* Ω :

$$a(\vec{u},\vec{v}) = \int_\Omega \frac{Ee}{1-\nu^2}\Big\{(1-\nu)\gamma^\alpha_\beta(\vec{u})\gamma^\beta_\alpha(\vec{v}) + \nu\gamma^\alpha_\alpha(\vec{u})\gamma^\beta_\beta(\vec{v}) + \frac{e^2}{12}\Big[(1-\nu)\bar{\rho}^\alpha_\beta(\vec{u})\bar{\rho}^\beta_\alpha(\vec{v}) + \nu\bar{\rho}^\alpha_\alpha(\vec{u})\bar{\rho}^\beta_\beta(\vec{v})\Big]\Big\}\sqrt{a}\, d\xi^1 d\xi^2 \tag{1.3.26}$$

<u>The work done by the external loads</u>

Now, it remains for us to indicate the expression of *the work done by the external loads*. We denote by

$$\vec{p} = p^i\vec{a}_i \tag{1.3.27}$$

the *resultant* on the middle surface S of the *surface* and *volume loads* applied to the shell $\mathcal{C}$. And, for simplicity, we assume that there are no *surface moments* and that the shell is *clamped* on a part of the boundary $\partial S_\circ$ of the middle surface. The corresponding part of the boundary Γ of the domain Ω is denoted $\Gamma_\circ$, i.e.,

$$\partial S_\circ = \phi(\Gamma_\circ)\ . \tag{1.3.28}$$

Then, the *work of the external loads* associated with a displacement $\vec{v} = v_i\vec{a}^i$ of the middle surface can be written

$$f(\vec{v}) = \int_\Omega p^i v_i \sqrt{a}\, d\xi^1 d\xi^2\ . \tag{1.3.29}$$

A somewhat different example will be given in Part II ; for more general boundary conditions, we refer to BERNADOU and CIARLET [1976] or KOITER [1966].

<u>1.4. Two equivalent formulations of the shell problem</u> :

The bilinear form a(.,.), given by (1.3.26), and the linear form f(.), given by (1.3.29), are defined on the plane open set Ω . The geometry of the surface S implicitly appears in the variable coefficients.

We can give sense to the expression (1.3.26) by taking $\vec{u},\vec{v} \in (H^1(\Omega))^2 \times H^2(\Omega)$, where $H^m(\Omega)$, $m = 1,2$, denote SOBOLEV spaces (see the introduction). In particular, for a sufficiently smooth mapping $\vec{\phi}$, we get,

$$\gamma^\alpha_\beta(\vec{v}) \in L^2(\Omega),\ \bar{\rho}^\alpha_\beta(\vec{v}) \in L^2(\Omega),\ \forall \vec{v} \in (H^1(\Omega))^2 \times H^2(\Omega), \qquad (1.4.1)$$

where the mixed components γ^α_β and $\bar{\rho}^\alpha_\beta$ are defined by relations (1.3.21) (1.3.22) and (1.3.25).

Then, adding the boundary conditions, we arrive at the space of *admissible displacements* $\vec{V}$:

$$\vec{V} = \{\vec{v} \,|\, \vec{v} \in (H^1(\Omega))^2 \times H^2(\Omega)\ ;\ \vec{v}|_{\Gamma_0} = \vec{0}\ ;\ \frac{\partial v_3}{\partial n}|_{\Gamma_0} = 0\} \qquad (1.4.2)$$

The space $\vec{V}$ is a HILBERT space when equipped with the scalar product induced by that of $(H^1(\Omega))^2 \times H^2(\Omega)$, that is,

$$((\vec{u},\vec{v})) = \sum_{\alpha=1}^{2} ((u_\alpha,v_\alpha))_{1,\Omega} + ((u_3,v_3))_{2,\Omega} \quad . \qquad (1.4.3)$$

The associated norm is denoted

$$\|\vec{v}\| = [((\vec{v},\vec{v}))]^{1/2} \quad . \qquad (1.4.4)$$

Then, we derive the following two equivalent formulations :

<u>Variational formulation of the problem</u> :

$$\left.\begin{array}{l} \textit{For } \vec{p} \in (L^2(\Omega))^3, \textit{ find } \vec{u} \in \vec{V} \textit{ such that} \\ a(\vec{u},\vec{v}) = f(\vec{v}) \ ,\ \forall \vec{v} \in \vec{V} \ , \end{array}\right\} \qquad (1.4.5)$$

or, alternatively, since the bilinear form a(.,.) is symmetric :

<u>Minimization formulation of the problem</u> :

$$\left.\begin{array}{l} \textit{For } \vec{p} \in (L^2(\Omega))^3, \textit{ find } \vec{u} \in \vec{V} \textit{ which minimizes the functional} \\ J : \vec{v} \in \vec{V} \to J(\vec{v}) = \frac{1}{2}\, a(\vec{v},\vec{v}) - f(\vec{v}) \,. \end{array}\right\} \qquad (1.4.6)$$

1.5. Other expressions for the bilinear form and the linear form :

With respect to the approximation of the solution of the continuous problem using finite element methods, it is convenient to use Theorems 1.5.1 and 1.5.2, which give other equivalent expressions for the bilinear form a(.,.) and for the linear form f(.).

Theorem 1.5.1 (CIARLET [1976, proposition 2.1]) :

The bilinear form a(.,.) *defined in* (1.3.26), *can be written*

$$a(\vec{u},\vec{v}) = \int_\Omega {}^t\mathbf{U}\,[\,\mathbf{A}_{IJ}]\,\mathbf{V}\,d\xi^1 d\xi^2, \tag{1.5.1}$$

where the column matrix $\mathbf{V}$ *(and similarly for the matrix* $\mathbf{U}$*) is given by*

$${}^t\mathbf{V} = [v_1\ v_{1,1}\ v_{1,2}\ v_2\ v_{2,1}\ v_{2,2}\ v_3\ v_{3,1}\ v_{3,2}\ v_{3,11}\ v_{3,12}\ v_{2,22}]\,. \tag{1.5.2}$$

Moreover, denoting by ϕ_i *the components of the mapping* $\vec{\phi} = \phi_i \vec{e}^i$ *on an orthonormal reference system* $(0,\ \vec{e}^1,\ \vec{e}^2,\ \vec{e}^3)$ *of the euclidean space* $\mathcal{E}^3$*, we have, for each* (I,J), $1 \le I,J \le 12$,

$$\mathbf{A}_{IJ}(\xi) = g_{IJ}(\phi_{i,\alpha}(\xi)\ ,\ \phi_{i,\alpha\beta}(\xi)\ ,\ \phi_{i,\alpha\beta\lambda}(\xi))\ , \tag{1.5.3}$$

where the function g_{IJ} *is a quotient of a polynomial in its arguments and a denominator which is an integer power of the expression*

$$\sqrt{a} = \sqrt{\det (a_{\alpha\beta})} = \left(\left(\sum_{i=1}^{3} (\phi_{i,1})^2\right) \left(\sum_{i=1}^{3} (\phi_{i,2})^2\right) - \left(\sum_{i=1}^{3} \phi_{i,1}\,\phi_{i,2}\right)^2 \right)^{\frac{1}{2}}\ . \tag{1.5.4}$$

∎

Let us indicate how to compute *explicitely* the coefficients $\mathbf{A}_{IJ}(\xi)$ which appear in the relation (1.5.1). We start from relation (1.3.26) and use relations (1.3.21) (1.3.22) and (1.3.25). With the notation (1.5.2), we get

$$\gamma^{\alpha}_{\beta}(\vec{v}) = \Lambda^{\alpha}_{\beta}\,\mathbf{V} \ , \tag{1.5.5}$$

where the four matrices Λ^{α}_{β}, of dimension (1,12), are given by

$$\left.\begin{aligned} \Lambda^{\alpha}_{\beta} = \Big[& -a^{\alpha\nu}\Gamma^{1}_{\beta\nu} \ ; \ a^{1\alpha}\delta_{\beta 1} \ ; \tfrac{1}{2}\,a^{\alpha\lambda}\Delta_{\lambda\beta} \ ; \ -\,a^{\alpha\nu}\Gamma^{2}_{\beta\nu} \ ; \\ & \tfrac{1}{2}\,a^{\alpha\lambda}\Delta_{\lambda\beta} \ ; \ a^{2\alpha}\delta_{\beta 2} \ ; \ -\,b^{\alpha}_{\beta} \ ; \ 0 \ ; \ 0 \ ; \ 0 \ ; \ 0 \ ; \ 0 \Big] \end{aligned}\right\} \tag{1.5.6}$$

with

$$\left.\begin{aligned} \delta_{11} = \delta_{22} = 1 \quad , \quad \delta_{12} = \delta_{21} = 0 \\ \Delta_{11} = \Delta_{22} = 0 \quad , \quad \Delta_{12} = \Delta_{21} = 1 \end{aligned}\right\} \tag{1.5.7}$$

Since $a^{\eta\alpha}|_{\beta} = 0$, we have

$$\left.\begin{aligned} \bar{\rho}^{\alpha}_{\beta}(\vec{v}) = a^{\alpha\eta}\bar{\rho}_{\eta\beta}(\vec{v}) = \ & a^{\alpha\eta}v_{3,\eta\beta} - a^{\alpha\eta}\Gamma^{\lambda}_{\eta\beta}v_{3,\lambda} - b^{\alpha\lambda}b_{\lambda\beta}v_{3} \\ & + (b^{\alpha\nu}|_{\beta} - b^{\alpha\lambda}\Gamma^{\nu}_{\lambda\beta} - b^{\lambda}_{\beta}a^{\alpha\eta}\Gamma^{\nu}_{\lambda\eta})v_{\nu} \ + \\ & + b^{\alpha\lambda}v_{\lambda,\beta} \ + b^{\lambda}_{\beta}a^{\alpha\eta}v_{\lambda,\eta} \ . \end{aligned}\right\}$$

Parallel to (1.5.5), we obtain

$$\bar{\rho}^{\alpha}_{\beta}(\vec{v}) = M^{\alpha}_{\beta}\,\mathbf{V} \tag{1.5.8}$$

with

$$\left.\begin{aligned} M^{\alpha}_{\beta} = \Big[& b^{1\alpha}|_{\beta} - b^{\alpha\lambda}\Gamma^{1}_{\lambda\beta} - a^{\alpha\nu}b^{\lambda}_{\beta}\Gamma^{1}_{\lambda\nu} \ ; \ b^{\alpha 1}\delta_{\beta 1} + a^{1\alpha}b^{1}_{\beta} \ ; \\ & b^{\alpha 1}\delta_{\beta 2} + a^{2\alpha}b^{1}_{\beta} \ ; \ b^{2\alpha}|_{\beta} - b^{\alpha\lambda}\Gamma^{2}_{\beta\lambda} - a^{\alpha\nu}b^{\lambda}_{\beta}\Gamma^{2}_{\lambda\nu} \ ; \\ & b^{\alpha 2}\delta_{\beta 1} + a^{1\alpha}b^{2}_{\beta} \ ; \ b^{\alpha 2}\delta_{\beta 2} + a^{2\alpha}b^{2}_{\beta} \ ; \ -\,b^{\alpha\lambda}b_{\lambda\beta} \ ; \\ & -\,a^{\alpha\lambda}\Gamma^{1}_{\lambda\beta} \ ; \ -\,a^{\alpha\lambda}\Gamma^{2}_{\lambda\beta} \ ; \ a^{1\alpha}\delta_{\beta 1} \ ; \ a^{\alpha\lambda}\Delta_{\lambda\beta} \ ; \ a^{2\alpha}\delta_{\beta 2} \Big] \end{aligned}\right\} \tag{1.5.9}$$

Coefficients $\delta_{\alpha\beta}$ and $\Delta_{\alpha\beta}$ are given by relations (1.5.7). Substituting relations (1.5.5) (1.5.8) into the relation (1.3.26) and comparing with (1.5.1), we obtain

$$\left.\begin{aligned}\left[\mathbf{A}_{IJ}\right] = \frac{Ee}{1-\nu^2}\sqrt{a}\ \Big\{ & (1-\nu)\,{}^t\Lambda^\alpha_\beta\,\Lambda^\beta_\alpha + \nu\,{}^t\Lambda^\alpha_\alpha\,\Lambda^\beta_\beta + \\ & + \frac{e^2}{12}(1-\nu)\,{}^tM^\alpha_\beta\,M^\beta_\alpha + \frac{e^2}{12}\nu\,{}^tM^\alpha_\alpha\,M^\beta_\beta \Big\} \ .\end{aligned}\right\} \quad (1.5.10)$$

In the numerical analysis to follow, we shall use numerical integration schemes. Therefore, we will have to evaluate the coefficients $\mathbf{A}_{IJ}$ at *finite* number of points of $\bar{\Omega}$. Assuming that $\vec{\phi} \in (\mathcal{C}^3(\bar{\Omega}))^3$, then $\mathbf{A}_{IJ} \in \mathcal{C}^\circ(\bar{\Omega})$. Hence, in order to compute the value of $\mathbf{A}_{IJ}(\xi)$ at point $\xi = (\xi^1,\xi^2) \in \bar{\Omega}$, we proceed as follows :

(1) Determine expressions as functions of ξ of the parameters

$$(a_{\alpha\beta}),\ a,\ (a^{\alpha\beta}),\ (b_{\alpha\beta}),\ (b^\alpha_\beta),\ (b^{\alpha\beta}),\ (c^\alpha_\beta = b^\alpha_\lambda b^\lambda_\beta), \Gamma^\alpha_{\beta\gamma},\ b^{\alpha\beta}|_\gamma \quad (1.5.11)$$

for all 33 functions (taking into account the symmetries when possible) ;

(2) Evaluate these 33 functions at the point ξ under consideration ;

(3) Compute Λ^α_β and M^α_β , and

(4) Deduce from the previous step, $\mathbf{A}_{IJ}(\xi) \in \mathbb{R}$, ξ fixed.

It remains to make explicit the expressions of parameters (1.5.11) as functions of partial derivatives of $\vec{\phi}$. We find

$$a_{\alpha\beta} = \vec{\phi}_{,\alpha}\cdot\vec{\phi}_{,\beta} \quad (1.5.12)$$

$$a = a_{11}a_{22} - (a_{12})^2 = (\vec{\phi}_{,1}\cdot\vec{\phi}_{,1})(\vec{\phi}_{,2}\cdot\vec{\phi}_{,2}) - (\vec{\phi}_{,1}\cdot\vec{\phi}_{,2})^2 \quad (1.5.13)$$

$$a^{11} = \frac{a_{22}}{a},\ a^{12} = a^{21} = -\frac{a_{12}}{a},\ a^{22} = \frac{a_{11}}{a}\ (\Leftrightarrow [a^{\alpha\beta}] = [a_{\alpha\beta}]^{-1}) \quad (1.5.14)$$

$$b_{\alpha\beta} = \frac{\vec{\phi}_{,1} \times \vec{\phi}_{,2}}{\sqrt{a}} \cdot \vec{\phi}_{,\alpha\beta}\ ,\ b^\alpha_\beta = a^{\alpha\lambda}b_{\lambda\beta}\ ,\ b^{\alpha\beta} = a^{\alpha\lambda}b^\beta_\lambda \quad (1.5.15)$$

$$\Gamma^\alpha_{\beta\gamma} = a^{\alpha\lambda}\vec{\phi}_{,\lambda}\cdot\vec{\phi}_{,\beta\gamma}\ , \quad (1.5.16)$$

$$c^{\alpha}_{\beta} = b^{\alpha}_{\lambda} b^{\lambda}_{\beta} , \tag{1.5.17}$$

$$b^{\alpha\beta}|_{\gamma} = a^{\alpha\lambda} a^{\beta\nu} b_{\lambda\nu|\gamma}$$

or equivalently,

$$\left. \begin{aligned} b^{\alpha\beta}|_{\gamma} = a^{\alpha\lambda} a^{\beta\nu} \Bigg\{ & \frac{\vec{\phi}_{,1} \times \vec{\phi}_{,2}}{\sqrt{a}} \cdot \vec{\phi}_{;\lambda\nu\gamma} \\ & -\Gamma^{\theta}_{\lambda\gamma} b_{\theta\nu} - \Gamma^{\theta}_{\gamma\nu} b_{\theta\lambda} - \Gamma^{\theta}_{\nu\lambda} b_{\theta\gamma} \Bigg\} \end{aligned} \right\} \tag{1.5.18}$$

Remark 1.5.1 : The assumption $\vec{\phi} \in (\mathcal{C}^3(\bar{\Omega}))^3$ can be weakened. For instance, in Part II, we will see an example for which the previous assumption is not satisfied. Nevertheless, it is possible to define the problem in the same way (see section 5.4).

∎

In a way similar to Theorem 1.5.1, we get

Theorem 1.5.2 :

The linear form (1.3.29) *can be written*

$$f(\vec{v}) = \int_{\Omega} {}^t\mathbf{F}\mathbf{V}\, d\xi^1 d\xi^2 \tag{1.5.19}$$

with

$${}^t\mathbf{F} = \sqrt{a}\,[p^1 \; 0 \; 0 \; p^2 \; 0 \; 0 \; p^3 \; 0 \; 0 \; 0 \; 0 \; 0] , \tag{1.5.20}$$

where the column matrix $\mathbf{V}$ *is given by the relation* (1.5.2).

∎

1.6. Existence and uniqueness of a solution :

In this section, we lay down conditions for the *existence and uniqueness of a solution* which are derived using ideas connected which mechanical origin of the problem, especially the so-called *infinitesimal rigid body lemma* recorded below. The absence of strains in rigid motions of shells is naturally viewed as a necessity by mechanicians for all physically reasonable shell theories. Proofs similar to that of Theorem 1.6.1 below also permit to establish conditions sufficient to guarantee the existence and the uniqueness of a solution for other types of shell equations. In particular, this kind of proof has been applied to the linear shell theories of NAGHDI [1963, 1972] (see COUTRIS [1976, 1978]).

Theorem 1.6.1 :

The variational problem (1.4.5) *(and thus, the equivalent minimization problem* (1.4.6)*) has one and only one solution.*

Proof (BERNADOU and CIARLET [1976])

We reiterate here the main points of the proof, which can be divided in five steps, involving six lemmas.

Step 1 : Equivalent norms

The definition of the bilinear form (1.3.26) uses general curvilinear coordinates and covariant derivatives. The following lemma generalizes the *KORN inequality* used in the two-dimensional elasticity problem (see, for example, DUVAUT and LIONS [1972], or HLAVÁČEK and NEČAS [1970]).

Lemma 1.6.1 : *Let* $\vec{v} = v_\alpha \vec{a}^\alpha + v_3 \vec{a}^3$. *The mapping*

$$\psi \; : \vec{v} \in (H^1(\Omega))^2 \times H^2(\Omega) \to \mathbb{R} \,, \tag{1.6.1}$$

defined by

$$\psi(\vec{v}) = \Big\{ |v_1|^2 + |v_2|^2 + |v_1|_1|^2 + |v_1|_2 + v_2|_1|^2 + |v_2|_2|^2 + |v_3|^2 + |v_3|_1|^2 + |v_3|_2|^2 + |v_3|_{11}|^2 + |v_3|_{12}|^2 + |v_3|_{22}|^2 \Big\}^{\frac{1}{2}} \tag{1.6.2}$$

is an equivalent norm to the usual norm (1.4.4).

■

Let us note that in (1.6.2), we mean by $|.|$ the norm of the space $L^2(\Omega)$.

Step 2 : The (infinitesimal) rigid body motion lemma

This lemma is essential for the proof of Theorem 1.6.1. For any $\vec{v} \in (H^1(\Omega))^2 \times H^2(\Omega)$, we denote

$$\Phi(\vec{v}) = \left\{ |\gamma_{11}(\vec{v})|^2 + |\gamma_{12}(\vec{v})|^2 + |\gamma_{22}(\vec{v})|^2 + |\bar{\rho}_{11}(\vec{v})|^2 + |\bar{\rho}_{12}(\vec{v})|^2 + |\bar{\rho}_{22}(\vec{v})|^2 \right\}^{1/2} , \tag{1.6.3}$$

where $|.|$ denotes the norm of the space $L^2(\Omega)$ and where the middle surface strain tensor $\gamma_{\alpha\beta}$ and the tensor of changes of curvature $\bar{\rho}_{\alpha\beta}$ are given by the relations (1.3.21) and (1.3.22), respectively.

Lemma 1.6.2 ((infinitesimal) rigid body motion lemma) : *Let us assume that the middle surface $\bar{S}$ is defined through a mapping $\vec{\phi} \in (\mathcal{C}^3(\bar{\Omega}))^3$. Then, we have the following equivalence*

$$\left(\Phi(\vec{v}) = 0, \; \vec{v} \in (H^1(\Omega))^2 \times H^2(\Omega) \right) \Leftrightarrow \left(\begin{array}{c} \vec{v} = \vec{A} + \vec{B} \times \vec{\phi} \\ \vec{A}, \vec{B} \text{ constant vectors} \\ \text{of } \mathcal{E}^3 \end{array} \right) .$$

■

Lemma 1.6.3 and 1.6.4 are immediate consequences of the (infinitesimal) rigid body motion lemma :

Lemma 1.6.3 : *Let $\vec{V}$ be the space (1.4.2) of the admissible displacements and let Φ be the functional (1.6.3). Then, we have the following equivalence*

$$\left(\Phi(\vec{v}) = 0, \; \vec{v} \in \vec{V} \right) \Leftrightarrow \vec{v} = \vec{0}.$$

■

Lemma 1.6.4 : *The functional Φ defined by (1.6.3) is a norm on the space $\vec{V}$ of the admissible displacements.*

■

Step 3 : The norms ψ and Φ are equivalent on the space $\vec{V}$.

Lemma 1.6.5 : *On the HILBERT space $\vec{V}$, the norms ψ and Φ , respectively defined by the relations (1.6.2) and (1.6.3), are equivalent.*

<u>Proof</u> :

(i) One readily checks the existence of a constant $K > 0$ such that

$$\Phi(\vec{v}) \leq K\psi(\vec{v}) \quad , \quad \forall \vec{v} \in \vec{V} \quad .$$

(ii) There exist two constants $L > 0$ and $M > 0$ such that

$$\left.\begin{aligned} &\psi^2(\vec{v}) \leq L\Phi^2(\vec{v}) + M(|v_1|^2 + |v_2|^2 + |v_3|^2) + \\ &\qquad + |v_3|_1|^2 + |v_3|_2|^2 \quad , \\ &\forall \vec{v} \in \vec{V} \quad . \end{aligned}\right\} \tag{1.6.4}$$

(iii) The map $\Phi : \vec{v} \in \vec{V} \to \Phi(\vec{v}) \in \mathbb{R}$ is weakly lower semi-continuous. Indeed, it is convex and (strongly) continuous on $\vec{V}$.

(iv) Any sequence $(\vec{v}_n)$ of elements of $\vec{V}$ such that, for all n,

$$\psi(\vec{v}_n) = 1 \quad , \quad \Phi(\vec{v}_n) < \frac{1}{n} \quad , \tag{1.6.5}$$

converges to $\vec{0}$, weakly in $\vec{V}$, strongly in $(L^2(\Omega))^2 \times H^1(\Omega)$.

(v) There exists $K > 0$, K constant, such that

$$\psi(\vec{v}) \leq K\Phi(\vec{v}) \quad , \quad \forall \vec{v} \in \vec{V} \quad . \tag{1.6.6}$$

Otherwise, there exists a sequence $(\vec{v}_n)$ satisfying the relations (1.6.5). Then, substituting $\vec{v}_n$ for $\vec{v}$ into relation (1.6.4), we get

$$1 \leq L\Phi^2(\vec{v}_n) + M(|v_{1n}|^2 + |v_{2n}|^2 + |v_{3n}|^2) + |v_{3n}|_1|^2 + |v_{3n}|_2|^2 \quad .$$

The results of (iv) and the inequality $\Phi(\vec{v}_n) < \frac{1}{n}$ involve the contradiction when $n \to +\infty$. Thus, the inequality (1.6.6) is true. ∎

<u>Step 4 : The map</u> $\vec{v} \in \vec{V} \to [a(\vec{v},\vec{v})]^{1/2}$ <u>is an equivalent norm to</u> $\|\vec{v}\|$.

One can easily check the existence of a constant $K > 0$ such that $a(\vec{v},\vec{v}) \geq K(\Phi(\vec{v}))^2$. Then, the Lemmas 1.6.1 and 1.6.5 lead to the next lemma :

Lemma 1.6.6 : *The bilinear form* a(.,.) *defined by* (1.3.26) *is* $\vec{V}$- *elliptic, that is, there exists a constant* $\alpha > 0$ *such that*

$$a(\vec{v},\vec{v}) \geq \alpha \|\vec{v}\|^2 \quad , \forall \vec{v} \in \vec{V} \quad . \tag{1.6.7}$$

■

Step 5 : Existence and uniqueness of a solution

From the results established thusfar, we conclude that the bilinear form a(.,.) is continuous, symmetric and $\vec{V}$- elliptic, and the linear form f(.) is continuous. Thus, the assertions of Theorem 1.6.1 now follow immediately from the *LAX and MILGRAM Theorem*.

■

CHAPTER 2

THE DISCRETE PROBLEM

Orientation :

The initial development of finite element analysis was largely due to the needs of aerospace structural design, where the theory of thin shells was used everyday. This fact explains the great number of papers devoted to this subject by engineers. A broad picture of much current thought and research on the application of different types of finite element methods to shell problems, with emphasis on numerical aspects, is done in ASHWELL-GALLAGHER [1976]. In this book, KNOWLES, RAZZAQUE and SPOONER [1976, p. 245] classify the finite element concepts for the representation of generally curved thin shells into three groups :

(i) The faceted form using *flat elements* ;

(ii) *Curved shell elements* formulated directly from appropriate thin shell theories ;

(iii) *Isoparametric solid elements* specialized to tackle thin shells by applying, in discrete form, appropriate thin shell assumptions (for example, Kirchhoff's normality hypothesis).

And these authors add : "For practical problems no one type of element has yet been found to be preeminent. The generally better performance of the higher-order elements cannot always be exploited and often is outweighed by their complexity in use. Conversely, the simplicity of the flat facet elements is all too often an enticement to use them in situations where sound performance cannot be guaranteed".

By contrast to the very numerous engineering studies on the approximation of general shell problems by finite element methods, there are very few papers that deal with the numerical analysis of such methods.

Related to the methods of group (i), i.e., *flat elements*, an analysis of the general problem is in progress by BERNADOU and DUCATEL. This study tries to generalize the results of CIARLET [1978, § 8.3] and JOHNSON [1975] who have considered the case of a circular arch problem and is based on a practical method described in CLOUGH and JOHNSON [1970].

Related to the methods of group (ii), i.e., *curved shell elements*, the first general study has been realized by CIARLET [1976]. Next, the effect of numerical integration and curved boundaries has been taken into account in BERNADOU [1978]. These studies are based on practical methods introduced by ARGYRIS and LOCHNER [1972], ARGYRIS, HAASE and MALEJANNAKIS [1973], DUPUIS and GOËL [1970a, 1970b], and DUPUIS [1971], among others. Also, the numerical analysis of *mixed* and *hybrid methods* for general shells are respectively considered by DESTUYNDER and LUTOBORSKI [1980] and STEPHAN and WEISSGERBER [1978].

Related to the methods of group (iii), i.e., *isoparametric solid elements*, it seems that the numerical analysis is not yet done. Practically, these methods have been first introduced by WEMPNER, ODEN, and KROSS [1968] and then studied by AHMAD, IRONS, and ZIENKIEWICZ [1970], ZIENKIEWICZ, TAYLOR and TOO [1971], and others.

In the following, we shall study a method of group (ii), i.e., *a conforming finite element method* with *numerical integration* for the general shell problem within the scope of the *linear* shell theory of W.T. KOITER. The treatment is essentially based on BERNADOU [1978, 1980].

We start in section 2.1 by defining the finite element space $\vec{V}_h$ which is a finite dimensional subspace of the space $\vec{V}$. Next, in section 2.2, we define the discrete problem, using integration schemes. Anticipating on the mathematical studies of section 2.4, we give in section 2.3 some examples of error estimates. These are obtained by using suitable integration schemes. The readers primarily interested in implementation and applications can skip section 2.4 and go directly to the next chapter.

We emphasize that, on the one hand, this study is applicable to *general linear shell equations* using any *system of curvilinear coordinates*, and that, on the other hand, this study provides *criteria* for the choice of suitable numerical integration schemes. This last feature is apparently new : up to now, the choice of numerical integration schemes seemed to be based on *empirical* considerations.

2.1. The finite element space $\vec{V}_h$:

From now on, we shall assume that the set $\bar{\Omega}$ is a *polygon*. Then, we may exactly cover the set $\bar{\Omega}$ by triangulations (composed of triangles or quadrilaterals). In the following, we consider a *regular family* of triangulations $\mathcal{T}_h$, i.e.,

(i) there exists a constant σ such that

$$\forall K \in \bigcup_h \mathcal{T}_h \ , \ \frac{h_K}{\rho_K} \leq \sigma \ , \tag{2.1.1}$$

where h_K = diam(K) and ρ_K = sup {diam(S), S is a ball contained in K} ;

(ii) the quantity

$$h = \max_{K \in \mathcal{T}_h} h_K \tag{2.1.2}$$

approaches zero.

In order to realize a *conforming approximation* of the solution $\vec{u}$ of the problem (1.4.5), we construct a finite dimensional *subspace* $\vec{V}_h$ of the space $\vec{V}$ defined in (1.4.2). First, we define a finite element subspace X_{h1} (resp. X_{h2}) of the space $H^1(\Omega)$ (resp. $H^2(\Omega)$) by using finite elements of classe $\mathcal{C}^\circ$ (resp. $\mathcal{C}^1$), i.e.,

$$X_{h1} \subset H^1(\Omega) \quad , \quad X_{h2} \subset H^2(\Omega) \quad . \tag{2.1.3}$$

Next, we consider the subspaces V_{h1} and V_{h2} of the spaces X_{h1} and X_{h2}, respectively defined by

$$V_{h1} = \{v_h \in X_{h1} \;;\; v_h = 0 \text{ on } \Gamma_\circ\} \;, \tag{2.1.4}$$

$$V_{h2} = \{v_h \in X_{h2} \;;\; v_h = \frac{\partial v_h}{\partial n} = 0 \text{ on } \Gamma_\circ\} \;. \tag{2.1.5}$$

In section 6.3, we show how to take into account boundary conditions such as $v_h = 0$ or $\partial_n v_h = 0$.

Then, we set

$$\vec{V}_h = V_{h1} \times V_{h1} \times V_{h2} \tag{2.1.6}$$

so that we have the inclusion

$$\vec{V}_h \subset \vec{V} \;. \tag{2.1.7}$$

When, (2.1.7) holds, we refer to the scheme as a *conforming finite element method.*

Examples : In order to construct the finite element space X_{h1}, we can use, for instance, the $\mathcal{C}^\circ$- finite elements described in the Figures 2.1.1 and 2.1.2. Similarly, to construct the finite element space X_{h2}, we can use the $\mathcal{C}^1$- finite elements described in the Figures 2.1.3 to 2.1.5.

In these figures, we denote by

* P_K the *function space of shape functions* for element K. In addition, we denote by $P_k(K)$ the space of all polynomials of degree $\leq k$;

* Σ_K the set of *degrees of freedom* ;

* a_i the vertices, b_i the midpoints of the sides of the elements.

The point of intersection of a line from a_i perpendicular to the opposite side is denoted c_i, as shown. The indices are counted modulo 3.

* Π_K the corresponding *interpolation operator*.

Moreover, the knowledge of the value of the function (resp. of its first differentials, resp. of its first and second differentials) at a point is indicated by a black point (resp. a circle surrounding this point, resp. two circles surrounding this point).

Remark 2.1.1 : Of course, the previous examples are not exhaustive. For instance, in order to define the space X_{h2} we could use some of the following $\mathcal{C}^1$- finite elements which are described in CIARLET [1978] :

(i) The triangle of BELL [1969] ;

(ii) The quadrilateral of FRAEIJS de VEUBEKE [1965] and SANDER [1969] ;

(iii) In case of a rectangular domain, the BOGNER, FOX, and SCHMIT [1965] rectangle.

Moreover, as space X_{h1}, we can take space X_{h2}. This choice is practically interesting because we only need to introduce one family of finite elements (of class $\mathcal{C}^1$). ∎

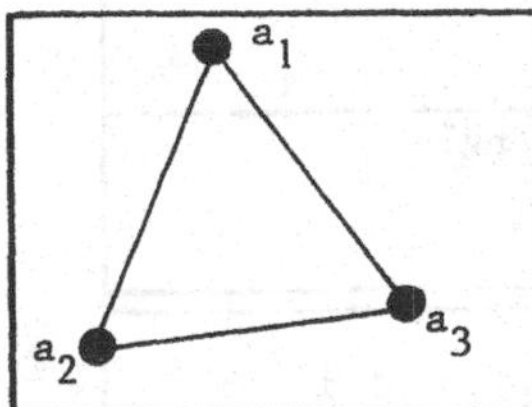

$P_K = P_1(K)$; dim. $P_K = 3$;

$\Sigma_K = \{p(a_i)\ ,\ 1 \le i \le 3\}$;

$\|v-\pi_K v\|_{m,K} \le Ch_K^{2-m}|v|_{2,K},\ \forall v \in H^2(K),\ 0 \le m \le 2$

Figure 2.1.1 : Triangle of type (1)

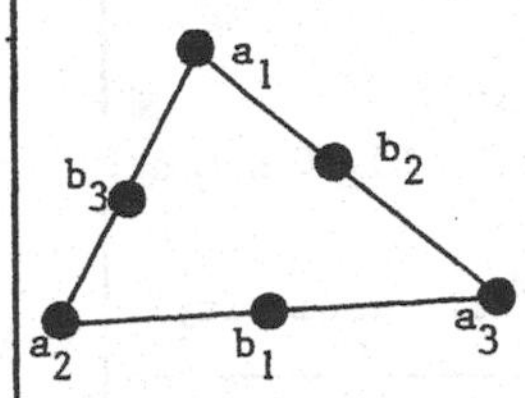

$P_K = P_2(K)$; dim. $P_K = 6$;

$\Sigma_K = \{p(a_i),\ p(b_i),\ 1 \le i \le 3\}$;

$\|v-\pi_K v\|_{m,K} \le Ch_K^{k+1-m}|v|_{k+1,K},\ \forall v \in H^{k+1}(K),$

$1 \le k \le 2$ et $0 \le m \le k+1$.

Figure 2.1.2 : Triangle of type (2)

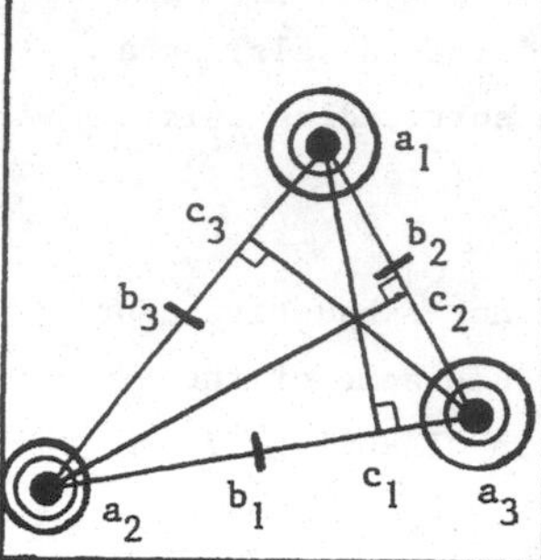

$P_K = P_5(K)$; $\dim P_K = 21$;

$\Sigma_K = \{p(a_i), Dp(a_i)(a_{i-1}-a_i), Dp(a_i)(a_{i+1}-a_i),$
$1 \le i \le 3,$
$D^2p(a_i)(a_{j+1}-a_{j-1})^2,\ 1 \le i,\ j \le 3,$
$Dp(b_i)(a_i-c_i),\ 1 \le i \le 3\}$

$\|v-\pi_K v\|_{m,K} \le Ch_K^{k+1-m}|v|_{k+1,K},\ \forall v \in H^{k+1}(K),$
$3 \le k \le 5$ et $0 \le m \le k+1$

Figure 2.1.3 : ARGYRIS' Triangle
(ARGYRIS, FRIED and SCHARPF [1968])

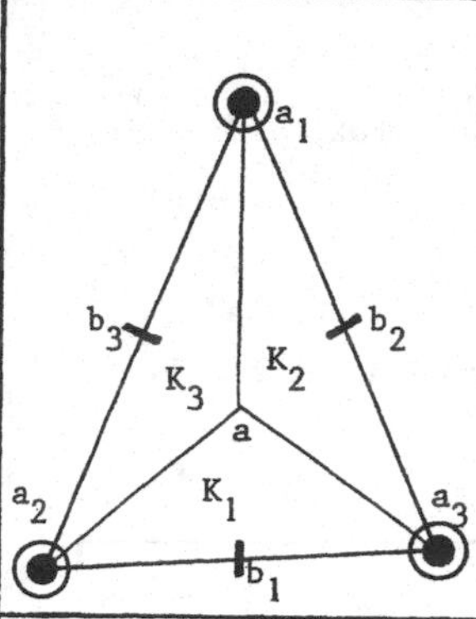

$P_K = \{p \in \mathcal{C}^1(K)$; $p|_{K_i} \in P_3(K_i),\ 1 \le i \le 3\}$;

$\dim P_K = 12$;

$\Sigma_K = \{p(a_i), Dp(a_i)(a_{i-1}-a_i), Dp(a_i)(a_{i+1}-a_i),$
$Dp(b_i)(a_i-c_i),\ 1 \le i \le 3\}$;

a = barycenter of triangle K (for simplicity)

$\|v-\pi_K v\|_{m,K} \le Ch_K^{k+1-m}|v|_{k+1,K},\ \forall v \in H^{k+1}(K),$
$k = 2,3$ et $0 \le m \le k+1$.

Figure 2.1.4 : HSIEH, CLOUGH and TOCHER Triangle
(CLOUGH and TOCHER [1965])

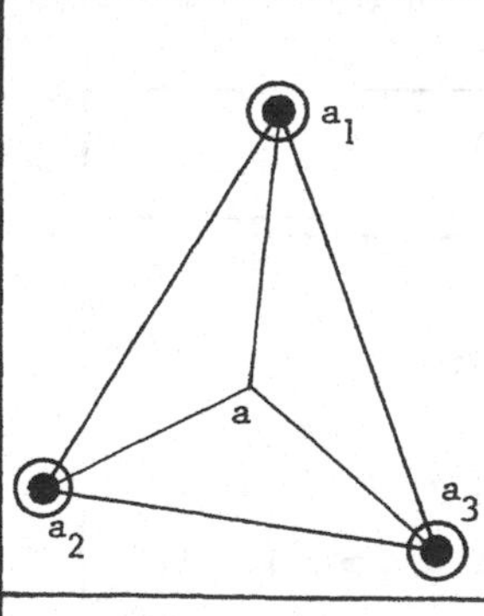

$P_K = \{p \in \mathcal{C}^1(K)$; $p|_{K_i} \in P_3(K_i),\ 1 \le i \le 3,$
$\partial_\nu p \in P_1(K')$ on each side K' of $K\}$;

$\dim P_K = 9$;

$\Sigma_K = \{p(a_i), Dp(a_i)(a_{i-1}-a_i), Dp(a_i)(a_{i+1}-a_i)$
$1 \le i \le 3\}$

$\|v-\pi_K v\|_{m,K} \le Ch_K^{3-m}|v|_{3,K},\ \forall v \in H^3(K),\ 0 \le m \le 3;$

a = barycenter of triangle K (for simplicity).

Figure 2.1.5 : Reduced HSIEH, CLOUGH and TOCHER Triangle

2.2. The discrete problem :

Taking into account the inclusion (2.1.7), i.e., $\vec{V}_h \subset \vec{V}$, a simple way to define a *discrete problem* associated to the continuous problem (1.4.5) can be : *Find* $\tilde{\vec{u}}_h \in \vec{V}_h$ *such that*

$$a(\tilde{\vec{u}}_h, \vec{v}_h) = f(\vec{v}_h) \quad , \forall \vec{v}_h \in \vec{V}_h \quad . \tag{2.2.1}$$

In particular, the inclusion $\vec{V}_h \subset \vec{V}$ ensures that the problem (2.2.1) has one and only one solution $\tilde{\vec{u}}_h \in \vec{V}_h$.

Nevertheless, an exact computation of the integrals which appear in (2.2.1) is very often impossible. In practice a *numerical integration scheme* is used, i.e., on every triangle $K \in \mathcal{T}_h$, we introduce the approximation

$$\int_K \phi(\xi^1,\xi^2) d\xi^1 d\xi^2 \sim \sum_{\ell=1}^{L} \omega_{\ell,K}\, \phi(b_{\ell,K}) \quad , \tag{2.2.2}$$

where $b_{\ell,K}$, $\omega_{\ell,K}$ denote respectively the *nodes* and the *weights* of the scheme. Thus, using (2.2.2), we are lead to solve the *new discrete problem*

Problem 2.2.1 : *Find* $\vec{u}_h \in \vec{V}_h$ *such that*

$$a_h(\vec{u}_h, \vec{v}_h) = f_h(\vec{v}_h) \quad , \quad \forall \vec{v}_h \in \vec{V}_h \; , \tag{2.2.3}$$

where (compare with relations (1.5.1) *and* (1.5.19), *respectively)* :

$$a_h(\vec{u}_h, \vec{v}_h) = \sum_{K \in \mathcal{T}_h} \sum_{\ell=1}^{L} \omega_{\ell,K} \; {}^t\mathbf{U}(b_{\ell,K}) [\mathbf{A}_{IJ}(b_{\ell,K})] \mathbf{V}_h(b_{\ell,K}) \; , \tag{2.2.4}$$

$$f_h(\vec{v}_h) = \sum_{K \in \mathcal{T}_h} \sum_{\ell=1}^{L} \omega_{\ell,K} \; {}^t\mathbf{F}(b_{\ell,K}) \mathbf{V}_h(b_{\ell,K}). \tag{2.2.5}$$

■

The hypothesis $\vec{\phi} \in \mathcal{C}^3(\bar{\Omega})$ implies $\mathbf{A}_{IJ} \in \mathcal{C}^\circ(\bar{\Omega})$ and thus $\mathbf{A}_{IJ}(b_{\ell,K})$ takes sense. Moreover, from now on we assume $\mathbf{F} \in (\mathcal{C}^\circ(\bar{\Omega}))^{12}$ – see (1.5.20) –. Then, we shall be interested in the following problems :

(i) *Show that the problem* (2.2.3) *has a unique solution* ;

(ii) *Determine sufficient conditions on numerical integration schemes to ensure that the error* $\|\vec{u} - \vec{u}_h\|$ *is of the same order as the* $\vec{V}_h$*-interpolation error, i.e.,*

$$\|\vec{u} - \vec{u}_h\| \sim \|\vec{u} - \overrightarrow{\Pi_h u}\| \tag{2.2.6}$$

where, in general, $\overrightarrow{\Pi_h v}$ *denotes the* $\vec{V}_h$*-interpolant of any function* $\vec{v} \in \vec{V}$.

These problems will be solved in section 2.4. Now, for clarity, we give some examples of numerical integration schemes which will be subsequently useful. For more details and examples of numerical integration schemes for *a triangle*, we refer for instance to COWPER [1973], HAASE [1977], HILLION [1977], LAURSEN and GELLERT [1978], and LYNESS and JESPERSEN [1975].

Examples of numerical integration schemes :

In the following examples, by S we mean the area of the triangle into consideration.

SCHEME 1 { HAMMER and STROUD [1956], STROUD [1971, page 307] ; Exact for polynomials of degree 2, 3 nodes.

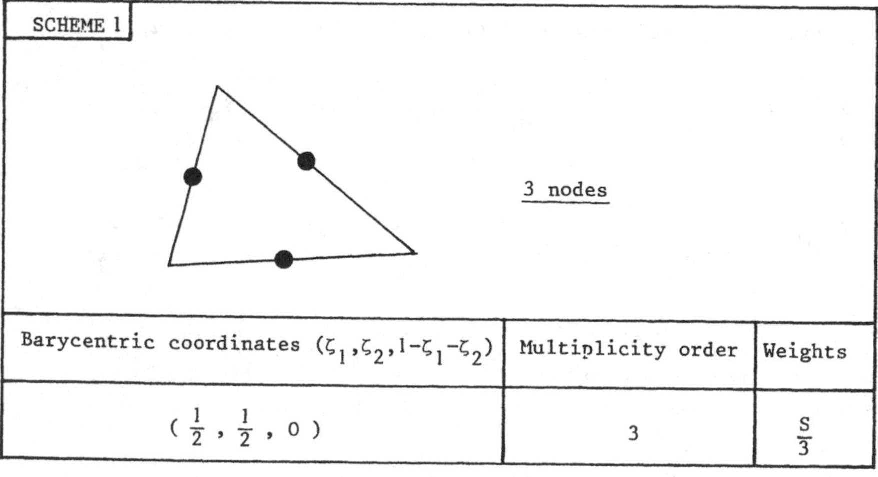

Barycentric coordinates $(\zeta_1, \zeta_2, 1-\zeta_1-\zeta_2)$	Multiplicity order	Weights
$(\frac{1}{2}, \frac{1}{2}, 0)$	3	$\frac{S}{3}$

Figure 2.2.1

SCHEME 2 — LYNESS and JESPERSEN [1975] ;
Exact for polynomials of order 4, 6 nodes.

SCHEME 2

6 nodes

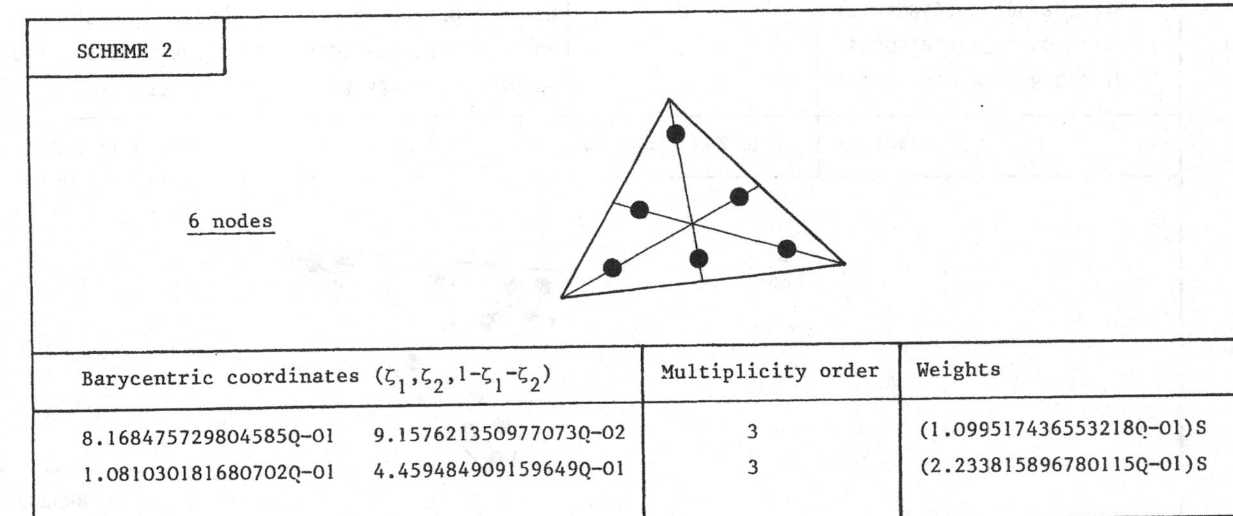

Barycentric coordinates $(\zeta_1, \zeta_2, 1-\zeta_1-\zeta_2)$		Multiplicity order	Weights
8.168475729804585Q-01	9.157621350977073Q-02	3	(1.099517436553218Q-01)S
1.081030181680702Q-01	4.459484909159649Q-01	3	(2.233815896780115Q-01)S

Figure 2.2.2

SCHEME 3 — LYNESS and JESPERSEN [1975] ; Exact for polynomials of order 8, 16 nodes .

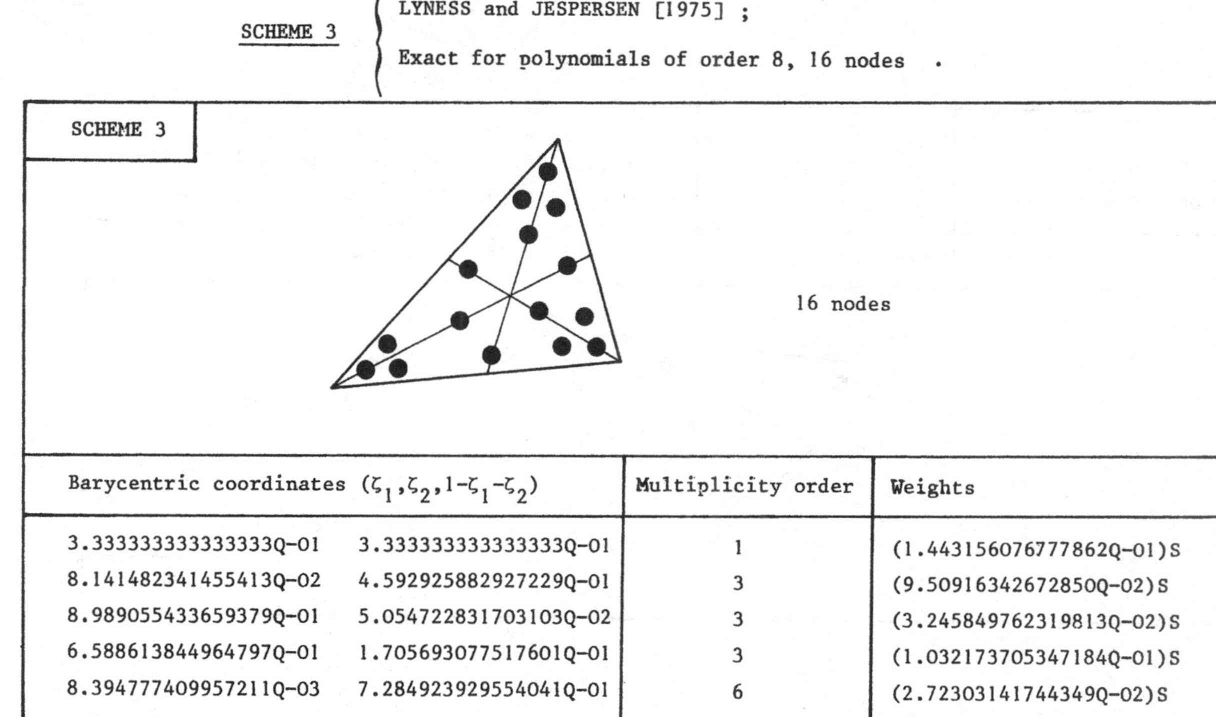

Barycentric coordinates $(\zeta_1,\zeta_2,1-\zeta_1-\zeta_2)$		Multiplicity order	Weights
3.333333333333333Q-01	3.333333333333333Q-01	1	(1.443156076777862Q-01)S
8.141482341455413Q-02	4.592925882927229Q-01	3	(9.509163426728500Q-02)S
8.989055433659379Q-01	5.054722831703103Q-02	3	(3.245849762319813Q-02)S
6.588613844964797Q-01	1.705693077517601Q-01	3	(1.032173705347184Q-01)S
8.394777409957211Q-03	7.284923929554041Q-01	6	(2.723031417443349Q-02)S

Figure 2.2.3

2.3. Examples of error estimates :

In order to have a presentation accessible to those primarily interested in the implementation and applications, we postpone the mathematical studies of convergence and error estimates to the next section. Here, we summarize results useful in the implementation of the method.

Thus, in Figure 2.3.1, we indicate in each case

(i) the result of the *error estimate* $\|\vec{u} - \vec{u}_h\|$ in the form $O(h^k)$;

(ii) the *hypothesis on the numerical scheme* ;

(iii) an *example* of a suitable numerical integration scheme ;

(iv) the *hypotheses on the solution* $\vec{u}$, *on the variable coefficients* A_{IJ} *and on the components* p^i *of the load* $\vec{p}$.

In all the cases the criterion on the choice of numerical integration schemes ensures the same asymptotic error estimate than in the case of exact integrations, for the same regularity assumption on the solution $\vec{u}$.

These results are given for some combinations of the spaces V_{h1} and V_{h2} which seem to be the best with respect to :

(i) the error estimate ;

(ii) the implementation : when the spaces V_{h1} and V_{h2} are associated to different types of finite elements, in order to define the space V_{h1} we use a finite element having the totality, or almost the totality, of its degrees of freedom included in the set of degrees of freedom of the finite element used to define the space V_{h2}.

In the case of HSIEH-CLOUGH-TOCHER triangles, the criterion on the numerical integration schemes has to be satisfied *on each subtriangle* K_i constituting the triangle. Practically, this observation is not of great concern in applications since the integrals on a composite

V_{h1} \ V_{h2}	ARGYRIS 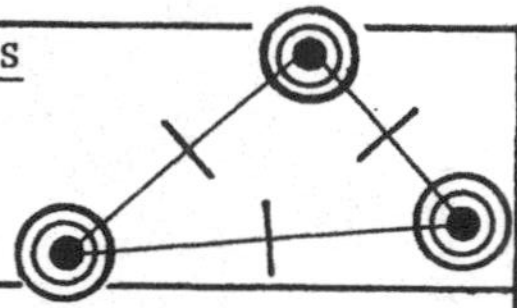	H.C.T. 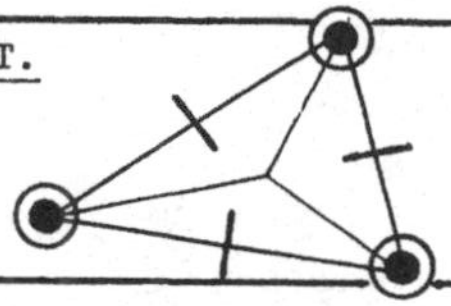	Reduced H.C.T. 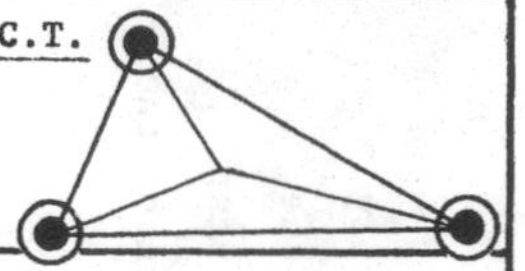
ARGYRIS	$O(h^4)$ Scheme exact for P_8 on K Scheme 3 Figure 2.2.3, 16 nodes $\vec{u}\|_K \in (H^5)^2 \times H^6$; $\mathbf{A}_{IJ} \in W^{4,\infty}$; $p^i \in W^{4,q}$, $q \geq 2$		
H.C.T.		$O(h^2)$ Scheme exact for P_4 on K_i Scheme 2, Figure 2.2.2, 6 nodes $\vec{u}\|_K \in (H^3)^2 \times H^4$; $\mathbf{A}_{IJ} \in W^{2,\infty}$; $p^i \in W^{2,q}$, $q \geq 2$	
Triangle of type (2) 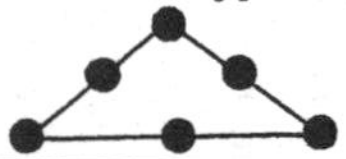		$O(h^2)$ Scheme exact for P_2 on K_i Scheme 1, Figure 2.2.1, 3 nodes $\vec{u}\|_K \in (H^3)^2 \times H^4$; $\mathbf{A}_{IJ} \in W^{2,\infty}$; $p^i \in W^{2,q}$, $q \geq 2$	
Reduced H.C.T.			$O(h)$ Scheme exact for P_4 on K_i Scheme 2, Figure 2.2.2, 6 nodes $\vec{u}\|_K \in (H^3)^3$; $\mathbf{A}_{IJ} \in W^{1,\infty}$; $p^i \in W^{1,q}$, $q>2$
Triangle of type (1) 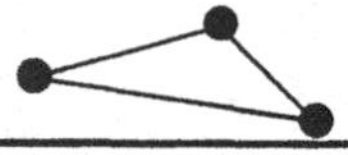			$O(h)$ Scheme exact for P_2 on K_i Scheme 1, Figure 2.2.1, 3 nodes $\vec{u}\|_K \in (H^2)^2 \times H^3$; $\mathbf{A}_{IJ} \in W^{1,\infty}$; $p^i \in W^{1,q}$, $q>2$

Figure 2.3.1 : Results of error estimates

(Successively in each case : (i) the error estimate $O(h^k)$; (ii) the hypothesis on the numerical integration schemes ; (iii) an example of a suitable scheme ; (iv) the regularity assumptions on the restrictions of $\vec{u}, \mathbf{A}_{IJ}$ and p^i on every triangle K (or subtriangle K_i in case of H.C.T. elements).

element $K = \bigcup_{i=1}^{3} K_i$ must be computed as a sum of integrals on each of the subtriangles K_i. Indeed, the interpolating functions change on each subtriangle.

Remark 2.3.1 : For the elements mentionned in Remark 2.1.1, the reader could similarly derive the corresponding error estimates from the results proved in section 2.4. These error estimates are explicitely given in BERNADOU [1978, 1980]. ∎

2.4. Mathematical studies of the convergence and of the error estimates :

Since the object of this book is to analyze applications of the finite element method to thin shell theory, it is appropriate for us to now detail, in a self-contrain way, the *mathematical theory* of the *convergence* and the method and the derivation of *error estimates* in the energy norm between the continuous and approximated solutions. Beyond the result of convergence, this study is very helpful because it leads to *criteria* for the choice of suitable numerical integration schemes (an apparently new feature since, up to now, the choice of numerical integration schemes seems to have been based on *empirical* considerations).

To develop criteria for convergence and to derive error estimates, it is useful to introduce the idea of an *affine regular family* of finite elements. This name encompasses two different notions : for simplicity, we discuss these notions for the case of triangular elements.

(i) *Reference triangle* $\hat{K}$: The domain $\bar{\Omega}$ is assumed to be covered by a *regular family* of triangulations $\mathcal{C}_h$, i.e., $\bar{\Omega} = \bigcup_{K\in\mathcal{C}_h} K$. Then, we call a triangle $\hat{K}$ *a reference triangle* if for each triangle $K \in \mathcal{C}_h$ there exists a unique invertible affine mapping

$$F_K : \hat{x} \in \hat{K} \to F_K(\hat{x}) = B_K\hat{x} + b_K \in K \quad , \tag{2.4.1}$$

where B_K is an invertible matrix and b_K is a vector of $\mathbb{R}^2$, such that

$$F_K(\hat{a}_i) = a_i \quad , \ i=1,2,3 \quad . \tag{2.4.2}$$

The points $\hat{a}_i, a_i$ are respective vertices of the triangles $\hat{K}$ and K.

(ii) *Reference finite element* $(\hat{K},\hat{P},\hat{\Sigma})$: we say that a *regular family of finite elements* (K,P_K,Σ_K), $K \in \mathcal{C}_h$, is *affine* if there exists a *reference finite element* $(\hat{K},\hat{P},\hat{\Sigma})$ such that the sets $\hat{K}$ and K, the spaces $\hat{P}$ and P_K, the sets of degrees of freedom $\hat{\Sigma}$ and Σ_K are in bijective correspondence through the invertible affine mapping F_K defined by (2.4.1) (2.4.2) for all $K \in \mathcal{C}_h$.

This notion of affine regular family of finite elements is convenient for obtaining interpolation properties. Even though most finite elements of class $\mathcal{C}^1$ do not form affine families, it was established by CIARLET [1978] that their interpolation properties are quite similar to those of affine families. Such families are said to be *almost-affine*. We will see that taking into account the numerical integration requires only the notion of the reference set $\hat{K}$, excluding the notion of reference finite element $(\hat{K},\hat{P},\hat{\Sigma})$, and therefore, *it can be applied to affine or almost affine regular families of finite elements indifferently.*

In this section 2.4, we denote the coordinates in the plane $\mathcal{E}^2$ by $x = (x_1,x_2)$ or $\hat{x} = (\hat{x}_1,\hat{x}_2)$ instead of the usual shell notation $\xi = (\xi^1,\xi^2)$ previously used. For any integer m, we let $P_m(K)$ denote the space of all the polynomials in x_1,x_2 of degree $\le m$, $(x_1,x_2) \in K$. Then, by analogy with CIARLET [1978, Theorem 8.2.4], one can prove the following theorem, in the case of *exact* integrations.

<u>Theorem 2.4.1</u> :

Let there be given two affine, or almost affine, regular families of finite elements (K,P_{K1},Σ_{K1}), (K,P_{K2},Σ_{K2}) *such that the corresponding space* $\vec{V}_h$ *satisfies the inclusion* (2.1.7). *Moreover, we assume that for all* $K \in \mathcal{C}_h$,

$$P_{m_1}(K) \subset P_{K1} \quad , \quad m_1 \in \mathbb{N} \quad , \quad 1 \le m_1 \quad , \tag{2.4.3}$$

$$P_{m_2}(K) \subset P_{K2} \quad , \quad m_2 \in \mathbb{N} \quad , \quad 2 \le m_2 \quad , \tag{2.4.4}$$

$$H^{m+1+\alpha}(K) \hookrightarrow \mathcal{C}^{s_\alpha}(K) \quad , \quad \alpha = 1,2 \quad , \quad m = -1 + \min(m_1,m_2-1) \quad , \tag{2.4.5}$$

where s_α *denotes the greatest order of partial derivatives occuring in the definition of the set* $\Sigma_{K\alpha}, \alpha = 1,2$.

Then, if the solution $\vec{u} = (u_1, u_2, u_3)$ *of the problem* (1.4.5) *belongs to the space* $(H^{m+2}(\Omega))^2 \times H^{m+3}(\Omega)$, *there exists a constant* C, *independent of* h, *such that*

$$\|\vec{u}-\tilde{\vec{u}}_h\| \le Ch^{m+1} \left(|u_1|^2_{m+2,\Omega} + |u_2|^2_{m+2,\Omega} + |u_3|^2_{m+3,\Omega}\right)^{1/2} , \qquad (2.4.6)$$

where $\tilde{\vec{u}}_h$ *is the approximate solution defined by* (2.2.1).

■

In section 2.2, we have emphasized that the discrete problem (2.2.1) is not realistic. Thus, we have essentially introduced a new discrete problem by using a *numerical integration technique*. For our purposes, it is convenient to introduce a *numerical integration scheme over the reference set* $\hat{K}$ according to

$$\int_{\hat{K}} \hat{\phi}(\hat{x})d\hat{x} \sim \sum_{\ell=1}^{L} \hat{\omega}_\ell \hat{\phi}(\hat{b}_\ell) . \qquad (2.4.7)$$

All the integrals appearing in (2.2.1) are of the form $\int_K \phi(x)dx$. Assuming, without loss of generality, that $\det(B_K) > 0$, we can write

$$\int_K \phi(x)dx = \det(B_K) \int_{\hat{K}} \hat{\phi}(\hat{x})d\hat{x} , \qquad (2.4.8)$$

using the usual correspondence between ϕ and $\hat{\phi}$, through the affine invertible mapping F_K defined in (2.4.1), i.e., $\hat{\phi} = \phi \circ F_K$ and $\phi = \hat{\phi} \circ F_K^{-1}$. Then, the numerical integration scheme (2.4.7) over the set $\hat{K}$ automatically induces a numerical integration scheme over the set K, namely

$$\int_K \phi(x)dx \sim \sum_{\ell=1}^{L} \omega_{\ell,K}\phi(b_{\ell,K}) , \qquad (2.4.9)$$

with

$$\omega_{\ell,K} = \det(B_K)\, \hat{\omega}_\ell \text{ and } b_{\ell,K} = F_K(\hat{b}_\ell) , \quad 1 \le \ell \le L . \qquad (2.4.10)$$

Moreover, we define the *error functionals*

$$\hat{E}(\hat{\phi}) = \int_{\hat{K}} \hat{\phi}(\hat{x})d\hat{x} - \sum_{\ell=1}^{L} \hat{\omega}_\ell \hat{\phi}(\hat{b}_\ell) , \qquad (2.4.11)$$

$$E_K(\phi) = \int_K \phi(x)dx - \sum_{\ell=1}^{L} \omega_{\ell,K}\phi(b_{\ell,K}) \tag{2.4.12}$$

so that

$$E_K(\phi) = \det(B_K)\ \hat{E}(\hat{\phi}) \quad . \tag{2.4.13}$$

The new discrete problem :

Thus, the *new discrete problem*, taking into account the use of numerical integration, is defined by (2.2.3), i.e. : *Find* $\vec{u}_h \in \vec{V}_h$ *such that*

$$a_h(\vec{u}_h,\vec{v}_h) = f_h(\vec{v}_h) \ , \ \forall \vec{v}_h \in \vec{V}_h \quad , \tag{2.4.14}$$

where $a_h(.,.)$ *and* $f_h(.)$ *are defined in* (2.2.4) (2.2.5), *i.e.,*

$$a_h(\vec{u}_h,\vec{v}_h) = \sum_{K\in\mathcal{C}_h} \sum_{\ell=1}^{L} \omega_{\ell,K}\ {}^t\mathbf{U}(b_{\ell,K})\left[\mathbf{A}_{IJ}(b_{\ell,K})\right]\mathbf{V}_h(b_{\ell,K}) \ , \tag{2.4.15}$$

$$f_h(\vec{v}_h) = \sum_{K\in\mathcal{C}_h} \sum_{\ell=1}^{L} \omega_{\ell,K}\ {}^t\mathbf{F}(b_{\ell,K})\ \mathbf{V}_h(b_{\ell,K}) \ . \tag{2.4.16}$$

The hypothesis $\vec{\phi} \in (\mathcal{C}^3(\bar{\Omega}))^3$ implies $\mathbf{A}_{IJ} \in \mathcal{C}^{\circ}(\bar{\Omega})$ and thus $\mathbf{A}_{IJ}(b_{\ell,K})$ makes sense. Moreover, from now on, we assume that $\mathbf{F} \in (\mathcal{C}^{\circ}(\bar{\Omega}))^{12}$ - see (1.5.20) -. Then, we shall be interested in the following problems :

(i) *Show that the problem* (2.4.14) *has a unique solution :* this will be achieved by showing that, under mild assumptions, the bilinear form $a_h(.,.)$ is $\vec{V}_h$-elliptic, uniformly with respect to h : see Theorem 2.4.5.

(ii) *Find sufficient conditions on numerical integration schemes which ensure that*

$$\|\vec{u}-\vec{u}_h\| = O(h^{m+1}) \quad , \tag{2.4.17}$$

i.e., the same order as in (2.4.6).

Remark 2.4.1 : The relations (2.4.15) (2.4.16) show the two different viewpoints between the CIARLET approximation [1978, Theorem 8.2.4] and this one :

(i) we take into account the numerical integration ;

(ii) the geometry of the shell appears only in the terms $\mathbf{A}_{IJ}(\xi)$ which are not approximated here, but only evaluated at the nodes of the numerical integration scheme. ∎

Abstract error estimate :

In order to solve the problem (2.4.17), we first give an "*abstract*" *error estimate*.

Theorem 2.4.2 :

Let us consider a family of discrete problems (2.4.14) *for which the bilinear forms are* $\vec{V}_h$*-elliptic, uniformly with respect to* h, *in the sense that there exists a constant* $\beta > 0$, *independent of* h, *such that :*

$$\beta\|\vec{v}_h\|^2 \leq a_h(\vec{v}_h,\vec{v}_h) \quad , \forall\vec{v}_h \in \vec{V}_h \quad . \tag{2.4.18}$$

Then, there exists a constant C, *independent of* h , *such that*

$$\|\vec{u}-\vec{u}_h\| \leq C \left\{ \inf_{\vec{v}_h\in\vec{V}_h} \left\{ \|\vec{u}-\vec{v}_h\| + \sup_{\vec{w}_h\in\vec{V}_h} \frac{|a(\vec{v}_h,\vec{w}_h)-a_h(\vec{v}_h,\vec{w}_h)|}{\|\vec{w}_h\|} \right\} + \sup_{\vec{w}_h\in\vec{V}_h} \frac{|f(\vec{w}_h)-f_h(\vec{w}_h)|}{\|\vec{w}_h\|} \right\} \tag{2.4.19}$$

Proof :

The assumption of $\vec{V}_h$-ellipticity involves the existence and the uniqueness of a solution $\vec{u}_h$ for the discrete problem (2.4.14). Then, let $\vec{v}_h$ be any element of the space $\vec{V}_h$. We are able to write

$$\left.\begin{aligned}\beta\|\vec{u}_h-\vec{v}_h\|^2 &\le a_h(\vec{u}_h-\vec{v}_h,\vec{u}_h-\vec{v}_h)\\ &= a(\vec{u}-\vec{v}_h,\vec{u}_h-\vec{v}_h) + [a(\vec{v}_h,\vec{u}_h-\vec{v}_h) - a_h(\vec{v}_h,\vec{u}_h-\vec{v}_h)] +\\ &\quad + [f_h(\vec{u}_h-\vec{v}_h) - f(\vec{u}_h-\vec{v}_h)] \quad ,\end{aligned}\right\}$$

so that the continuity of the bilinear form a(.,.) implies

$$\begin{aligned}\beta\|\vec{u}_h-\vec{v}_h\| &\le M\|\vec{u}-\vec{v}_h\| + \frac{|a(\vec{v}_h,\vec{u}_h-\vec{v}_h) - a_h(\vec{v}_h,\vec{u}_h-\vec{v}_h)|}{\|\vec{u}_h-\vec{v}_h\|} +\\ &\quad + \frac{|f_h(\vec{u}_h-\vec{v}_h) - f(\vec{u}_h-\vec{v}_h)|}{\|\vec{u}_h-\vec{v}_h\|}\\ &\le M\|\vec{u}-\vec{v}_h\| + \sup_{\vec{w}_h\in\vec{V}_h} \frac{|a(\vec{v}_h,\vec{w}_h) - a_h(\vec{v}_h,\vec{w}_h)|}{\|\vec{w}_h\|} +\\ &\quad + \sup_{\vec{w}_h\in\vec{V}_h} \frac{|f_h(\vec{w}_h) - f(\vec{w}_h)|}{\|\vec{w}_h\|} \quad .\end{aligned}$$

Combining this inequality with the triangular inequality

$$\|\vec{u}-\vec{u}_h\| \le \|\vec{u}-\vec{v}_h\| + \|\vec{v}_h-\vec{u}_h\| \quad ,$$

and taking the minimum with respect to $\vec{v}_h \in \vec{V}_h$, we get the inequality (2.4.19).

∎

Thus, in addition to the usual approximation theory term $\inf_{\vec{v}_h\in\vec{V}_h} \|\vec{u}-\vec{v}_h\|$, we find two additional terms which measure the *consistency* of the integration schemes for the bilinear form a(.,.) and for the linear form f(.), respectively.

To find an explicit estimate of the error, we generalize some results of CIARLET-RAVIART [unpublished] and BERNADOU-DUCATEL [1978] related to the approximation of problems of order 2 and order 4, respectively. In the Theorems 2.4.3 and 2.4.4, we prove local error estimates. Next, in the Theorem 2.4.5, we give sufficient conditions on the numerical integration schemes in order to get the uniform $\vec{V}_h$-ellipticity condition (2.4.18). Finally, in the Theorem 2.4.6, we derive the explicit error estimate.

Local error estimates :

We start by recalling the BRAMBLE-HILBERT lemma which is subsequently very useful :

Lemma 2.4.1 (BRAMBLE-HILBERT [1970]) : *Let Ω be a bounded open subset of $\mathbb{R}^n$ with a LIPSCHITZ-continuous boundary. For some integer $k \geq 0$ and some real number $p \in [1,+\infty]$, let f be a continuous linear form on the space $W^{k+1,p}(\Omega)$ such that*

$$\forall q \in P_k(\Omega) \ , \ f(q) = 0.$$

Then, there exists a constant $C(\Omega)$ such that

$$\forall v \in W^{k+1,p}(\Omega) \ , \ |f(v)| \ \leq C(\Omega) \ \|f\|^*_{k+1,p,\Omega} |v|_{k+1,p,\Omega} \ ,$$

where $\|\cdot\|^*_{k+1,p,\Omega}$ is the norm in the dual space of $W^{k+1,p}(\Omega)$. ∎

Theorem 2.4.3 :

Let us consider two finite element families (K,P_{K1},Σ_{K1}) and (K,P_{K2},Σ_{K2}) for which the triangles K are in correspondence with a reference triangle $\hat{K}$ through an invertible affine mapping F_K of type (2.4.1) and such that the finite dimensional spaces P_{K1} and P_{K2} verify the inclusions

$$P_{m_1}(K) \subset P_{K1} \subset P_{n_1}(K) \quad , \ 1 \leq m_1 \leq n_1 \quad , \qquad (2.4.20)$$

$$P_{m_2}(K) \subset P_{K2} \subset P_{n_2}(K) \quad , \ 2 \leq m_2 \leq n_2 \quad , \qquad (2.4.21)$$

the quantities m_1,n_1,m_2,n_2 denoting integer numbers. Let k,ℓ,m be integers ≥ 0. Let us assume that the numerical integration scheme on the reference triangle satisfies the following properties (the parameter α

(resp. β) takes value 1 *or* 2 *according to whether* $v \in P_{K\alpha} = P_{K1}$ *or* P_{K2} *(resp.* $w \in P_{K\beta} = P_{K1}$ or P_{K2}*))* :

(i) *If* $k+\ell-|\nu|-|\mu| \geq m+1$

$$\forall \hat{\phi} \in P_{m+n_\alpha-k}(\hat{K}) \quad , \hat{E}(\hat{\phi}) = 0 \quad \text{if } k-|\nu| \leq m \quad , \tag{2.4.22}$$

$$\forall \hat{\phi} \in P_{m+n_\beta-\ell}(\hat{K}) \quad , \hat{E}(\hat{\phi}) = 0 \quad \text{if } \ell-|\mu| \leq m \quad , \tag{2.4.23}$$

$$\forall \hat{\phi} \in P_m(\hat{K}) \quad , \quad \hat{E}(\hat{\phi}) = 0. \tag{2.4.24}$$

(ii) *If* $|\nu| = \alpha = k, |\mu| = \beta = \ell$ and $m=0$,

$$\forall \hat{\phi} \in P_{n_\alpha+n_\beta-\alpha-\beta}(\hat{K}) \quad , \quad \hat{E}(\hat{\phi}) = 0. \tag{2.4.25}$$

Then, there exists a constant $C > 0$*, independent of* K *, such that for all function* $b \in W^{m+1,\infty}(K)$*, for all* $v \in P_{K\alpha}$*,* $\alpha = 1$ *or* 2*, for all* $w \in P_{K\beta}$*,* $\beta = 1$ *or* 2*, for all* $\nu = (\nu_1,\nu_2)$ *verifying* $0 \leq |\nu| \leq \alpha \leq k$*, for all* $\mu = (\mu_1,\mu_2)$ *verifying* $0 \leq |\mu| \leq \beta \leq \ell$*, we have the estimate*

$$|E_K(b\partial^\nu v\partial^\mu w)| \leq Ch_K^{m+1} \|b\|_{m+1,\infty,K} \|v\|_{k,K} \|w\|_{\ell,K} \quad , \tag{2.4.26}$$

where $h_K = \text{diam}\,(K)$. ∎

<u>Proof</u> : (i) <u>If</u> $k+\ell-|\nu|-|\mu| \geq m+1$

Let us set $p = \partial^\nu v$ and $q = \partial^\mu w$. The inclusions (2.4.20) and (2.4.21) imply immediately

$$p \in P_{n_\alpha-|\nu|}(K), \; q \in P_{n_\beta-|\mu|}(K). \tag{2.4.27}$$

Since the map $F_K : \hat{K} \to K$ is affine, the relation (2.4.13) implies

$$E_K(b\partial^\nu v\partial^\mu w) = E_K(bpq) = \det(B_K)\hat{E}(\hat{b}\hat{p}\hat{q}) \tag{2.4.28}$$

with

$$\hat{b} \in W^{m+1,\infty}(\hat{K}) \quad , \; \hat{p} \in P_{n_\alpha-|\nu|}(\hat{K}) \quad , \; \hat{q} \in P_{n_\beta-|\mu|}(\hat{K}) \; . \tag{2.4.29}$$

By hypothesis $|\nu| \leq k$, $|\mu| \leq \ell$. Then, we can write

$$\left.\begin{aligned}\hat{E}(\hat{b}\hat{p}\hat{q}) = {} & \hat{E}[\hat{b}(\hat{\pi}_{k-|\nu|}\hat{p})(\hat{\pi}_{\ell-|\mu|}\hat{q})] + \hat{E}[\hat{b}(\hat{\pi}_{k-|\nu|}\hat{p})(\hat{q}-\hat{\pi}_{\ell-|\mu|}\hat{q})] \\ & + \hat{E}[\hat{b}(\hat{p}-\hat{\pi}_{k-|\nu|}\hat{p})(\hat{\pi}_{\ell-|\mu|}\hat{q})] + \\ & + \hat{E}[\hat{b}(\hat{p}-\hat{\pi}_{k-|\nu|}\hat{p})(\hat{q}-\hat{\pi}_{\ell-|\mu|}\hat{q})] ,\end{aligned}\right\} \quad (2.4.30)$$

where we generally denote by $\hat{\pi}_\ell$ the orthogonal projector in the HILBERT space $L^2(\hat{K})$ on the closed subspace $P_\ell(\hat{K})$. To simplify the arguments, we set

$$k_1 = k-|\nu| \ , \ \ell_1 = \ell-|\mu| \quad . \qquad (2.4.31)$$

Then, we consider the following four steps :

<u>Step 1</u> : <u>Estimate of</u> $\hat{E}(\hat{b}\hat{\pi}_{k_1}\hat{p}\hat{\pi}_{\ell_1}\hat{q})$, $\forall \hat{b} \in W^{m+1,\infty}(\hat{K})$, $\forall \hat{p} \in P_{n_\alpha-|\nu|}$, $\forall \hat{q} \in P_{n_\beta-|\mu|}$

The assumption (2.4.24) and Lemma 2.4.1 involve $|\hat{E}(\hat{\psi})| \leq C|\hat{\psi}|_{m+1,\infty,\hat{K}}$, $\forall \hat{\psi} \in W^{m+1,\infty}(\hat{K})$, where, from now on, the letters C denote constants which are independent of the triangle into consideration and which can change from one inequality to the other. Let us substitute $\hat{\psi}$ for $\hat{b}\hat{\pi}_{k_1}\hat{p}\hat{\pi}_{\ell_1}\hat{q}$. The LEIBNIZ formula gives

$$|\hat{E}(\hat{b}\hat{\pi}_{k_1}\hat{p}\hat{\pi}_{\ell_1}\hat{q})| \leq C \sum_{\substack{i+j=0 \\ i\leq k_1 \\ j\leq \ell_1}}^{m+1} |\hat{b}|_{m+1-i-j,\infty,\hat{K}} |\hat{\pi}_{k_1}\hat{p}|_{i,\infty,\hat{K}} |\hat{\pi}_{\ell_1}\hat{q}|_{j,\infty,\hat{K}} \ .$$

But the equivalence of norms on the finite dimensional spaces $P_{k_1}(\hat{K})$ and $P_{\ell_1}(\hat{K})$ involve $|\hat{\pi}_{k_1}\hat{p}|_{i,\infty,\hat{K}} \leq C|\hat{\pi}_{k_1}\hat{p}|_{i,\hat{K}}$ and $|\hat{\pi}_{\ell_1}\hat{q}|_{j,\infty,\hat{K}} \leq$ $\leq C|\hat{\pi}_{\ell_1}\hat{q}|_{j,\hat{K}}$. Since $\hat{\pi}_{k_1}$ is an orthogonal projector in $L^2(\hat{K})$, we have on the one hand $|\hat{\pi}_{k_1}\hat{p}|_{0,\hat{K}} \leq |\hat{p}|_{0,\hat{K}}$; on the other hand, the operator $\hat{\pi}_{k_1}$ leaves invariant the space $P_{i-1}(\hat{K})$, $1 \leq i \leq k_1$, so that the interpolation theorem gives

$$|\hat{p}-\hat{\pi}_{k_1}\hat{p}|_{i,\hat{K}} \leq C|\hat{p}|_{i,\hat{K}} \quad ,$$

and hence,

$$|\hat{\pi}_{k_1}\hat{p}|_{i,\hat{K}} \le |\hat{p}-\hat{\pi}_{k_1}\hat{p}|_{i,\hat{K}} + |\hat{p}|_{i,\hat{K}} \le C|\hat{p}|_{i,\hat{K}} \quad , \; 1 \le i \le k_1 \quad .$$

Finally, we get

$$|\hat{\pi}_{k_1}\hat{p}|_{i,\hat{K}} \le C|\hat{p}|_{i,\hat{K}} \quad , \; i = 0,\dots,k_1 \quad , \tag{2.4.32}$$

and a similar inequality for $|\hat{\pi}_{\ell_1}\hat{q}|_{j,\hat{K}}$. Then, combining all the previous inequalities, we arrive at the existence of a constant C such that

$$\left.\begin{aligned} &|\hat{E}(\hat{b}\hat{\pi}_{k_1}\hat{p}\hat{\pi}_{\ell_1}\hat{q})| \le C \sum_{\substack{i+j=0\\ i\le k_1\\ j\le \ell_1}}^{m+1} |\hat{b}|_{m+1-i-j,\infty,\hat{K}}|\hat{p}|_{i,\hat{K}}|\hat{q}|_{j,\hat{K}} \quad , \\ &\forall\hat{b}\in W^{m+1,\infty}(\hat{K}) \; , \; \forall\hat{p}\in P_{n_\alpha-|\nu|} \quad , \; \forall\hat{q}\in P_{n_\beta-|\mu|} \quad . \end{aligned}\right\} \tag{2.4.33}$$

<u>Step 2</u> : <u>Estimate of</u> $\hat{E}(\hat{b}\hat{\pi}_{k_1}\hat{p}(\hat{q}-\hat{\pi}_{\ell_1}\hat{q}))$ <u>and</u> $\hat{E}(\hat{b}(\hat{p}-\hat{\pi}_{k_1}\hat{p})\hat{\pi}_{\ell_1}\hat{q})$,

$$\forall\hat{b}\in W^{m+1,\infty}(\hat{K}) \; , \; \forall\hat{p}\in P_{n_\alpha-|\nu|} \; , \; \forall\hat{q}\in P_{n_\beta-|\mu|}$$

We start with the first of these two terms.

<u>If</u> $0 \le \ell_1 \le m$, we get the following inclusions (see the SOBOLEV's Embedding Theorem in the introduction of part I)

$$W^{m+1,\infty}(\hat{K}) \subsetneq W^{m+1-\ell_1,r}(\hat{K}) \subsetneq \mathcal{C}^{\circ}(\hat{K}) \quad , \tag{2.4.34}$$

with $r \ge 1$ and $r > \frac{2}{m+1-\ell_1}$. Particularly, from $W^{\ell_1,\infty}(\hat{K}) \subsetneq L^r(\hat{K})$, we derive

$$|\hat{\phi}|_{0,r,\hat{K}} \le C \sum_{j=0}^{\ell_1} |\hat{\phi}|_{\ell_1-j,\infty,\hat{K}} \quad , \; \forall\hat{\phi}\in W^{\ell_1,\infty}(\hat{K}) \quad ,$$

and hence,

$$|\hat{\phi}|_{m+1-\ell_1,r,\hat{K}} \le C \sum_{j=0}^{\ell_1} |\hat{\phi}|_{m+1-j,\infty,\hat{K}} \quad , \; \forall\hat{\phi}\in W^{m+1,\infty}(\hat{K}) \tag{2.4.35}.$$

By analogy with the arguments of the step 1, we find that for all

$\hat\phi \in W^{m+1-\ell_1,r}(\hat K) \subsetneq \mathcal{C}^\circ(\hat K)$, for all $\hat q \in P_{n_\beta-|\mu|}$,

$$|\hat E(\hat\phi(\hat q-\hat\pi_{\ell_1}\hat q))| \le C|\hat\phi(\hat q-\hat\pi_{\ell_1}\hat q)|_{0,\infty,\hat K} \le C|\hat\phi|_{0,\infty,\hat K}|\hat q-\hat\pi_{\ell_1}\hat q|_{0,\infty,\hat K}$$

$$\le C\|\hat\phi\|_{m+1-\ell_1,r,\hat K}|\hat q-\hat\pi_{\ell_1}\hat q|_{0,\hat K} \quad .$$

For a given $\hat q \in P_{n_\beta-|\mu|}$, the linear form

$$\hat\phi \in W^{m+1-\ell_1,r}(\hat K) \to \hat E(\hat\phi(\hat q-\hat\pi_{\ell_1}\hat q))$$

is continuous and is zero on the space $P_{m-\ell_1}(\hat K)$, thanks to the assumptions (2.4.23)(2.4.24) and (2.4.31) (note that the assumption $\ell_1 \le m$ and relations (2.4.31) imply $\ell-|\mu| \le m$). With the Lemma 2.4.1, we derive

$$\left.\begin{aligned} &|\hat E(\hat\phi(\hat q-\hat\pi_{\ell_1}\hat q))| \le C|\hat\phi|_{m+1-\ell_1,r,\hat K}|\hat q-\hat\pi_{\ell_1}\hat q|_{0,\hat K} \\ &\forall\hat\phi \in W^{m+1-\ell_1,r}(\hat K), \quad \forall\hat q \in P_{n_\beta-|\mu|} \quad . \end{aligned}\right\} \tag{2.4.36}$$

Since the operator $\hat\pi_{\ell_1}$ particularly leaves invariant the space $P_{j-1}(\hat K)$, $j = 1,\ldots,\ell_1$, we get

$$|\hat q-\hat\pi_{\ell_1}\hat q|_{0,\hat K} \le C|\hat q|_{j,\hat K} \quad , 0 \le j \le \ell_1, \quad \forall\hat q \in P_{n_\beta-|\mu|} \tag{2.4.37}$$

(for $j = 0$, we use the property of the projection in $L^2(\hat K)$). Then the relations (2.4.34) and the inequalities (2.4.35) to (2.4.37) imply

$$\left.\begin{aligned} &|\hat E(\hat\phi(\hat q-\hat\pi_{\ell_1}\hat q))| \le C \sum_{j=0}^{\ell_1} |\hat\phi|_{m+1-j,\infty,\hat K}|\hat q|_{j,\hat K} \ , \ \forall\hat\phi \in W^{m+1,\infty}(\hat K), \\ &\forall\hat q \in P_{n_\beta-|\mu|} \quad . \end{aligned}\right\}$$

Taking $\hat\phi = \hat b\hat\pi_{k_1}\hat p$, using the LEIBNIZ formula and inequalities (2.4.32), we get

$$\left.\begin{aligned}&|\hat{E}(\hat{b}\hat{\pi}_{k_1}\hat{p}(\hat{q}-\hat{\pi}_{\ell_1}\hat{q}))| \le C \sum_{\substack{i+j=0\\ i\le k_1\\ j\le \ell_1}}^{m+1} |\hat{b}|_{m+1-i-j,\infty,\hat{K}}|\hat{p}|_{i,\hat{K}}|\hat{q}|_{j,\hat{K}}\\ &\forall\hat{b}\in W^{m+1,\infty}(\hat{K}),\quad \forall\hat{p}\in P_{n_\alpha-|\nu|},\ \forall\hat{q}\in P_{n_\beta-|\mu|}\ .\end{aligned}\right\}\quad(2.4.38)$$

<u>If</u> $m+1\le \ell_1$, then $W^{m+1,\infty}(\hat{K})\subsetneq \mathcal{C}^\circ(\hat{K})$ implies

$$|\hat{E}(\hat{\phi}(\hat{q}-\hat{\pi}_{\ell_1}\hat{q}))| \le C|\hat{\phi}|_{0,\infty,\hat{K}}|\hat{q}|_{\ell_1,\hat{K}}\ ,\ \forall\hat{\phi}\in W^{m+1,\infty}(\hat{K})\ ,\ \forall\hat{q}\in P_{n_\beta-|\mu|}\ .$$

Taking $\hat{\phi}=\hat{b}\hat{\pi}_{k_1}\hat{p}$, we derive

$$\left.\begin{aligned}&|\hat{E}(\hat{b}\hat{\pi}_{k_1}\hat{p}(\hat{q}-\hat{\pi}_{\ell_1}\hat{q}))| \le C|\hat{b}|_{0,\infty,\hat{K}}|\hat{p}|_{0,\hat{K}}|\hat{q}|_{\ell_1,\hat{K}}\ ,\ \forall\hat{b}\in W^{m+1,\infty}(\hat{K}),\\ &\forall\hat{p}\in P_{n_\alpha-|\nu|}\quad ,\ \forall\hat{q}\in P_{n_\beta-|\mu|}\quad .\end{aligned}\right\}\quad(2.4.39)$$

In the same way, we obtain the following bounds for the second term :

<u>If</u> $0\le k_1\le m$ (which implies $k-|\nu|\le m$),

$$\left.\begin{aligned}&|\hat{E}(\hat{b}(\hat{p}-\hat{\pi}_{k_1}\hat{p})\hat{\pi}_{\ell_1}\hat{q}| \le C \sum_{\substack{i+j=0\\ i\le k_1\\ j\le \ell_1}}^{m+1} |\hat{b}|_{m+1-i-j,\infty,\hat{K}}|\hat{p}|_{i,\hat{K}}|\hat{q}|_{j,\hat{K}}\\ &\forall\hat{b}\in W^{m+1,\infty}(\hat{K}),\quad \forall\hat{p}\in P_{n_\alpha-|\nu|}\quad ,\ \forall\hat{q}\in P_{n_\beta-|\mu|}\quad .\end{aligned}\right\}\quad(2.4.40)$$

<u>If</u> $m+1\le k_1$

$$\left.\begin{aligned}&|\hat{E}(\hat{b}(\hat{p}-\hat{\pi}_{k_1}\hat{p})\hat{\pi}_{\ell_1}\hat{q})| \le C|\hat{b}|_{0,\infty,\hat{K}}|\hat{p}|_{k_1,\hat{K}}|\hat{q}|_{0,\hat{K}}\ ,\ \forall\hat{b}\in W^{m+1,\infty}(\hat{K})\ ,\\ &\forall\hat{p}\in P_{n_\alpha-|\nu|}\quad ,\ \forall\hat{q}\in P_{n_\beta-|\mu|}\quad .\end{aligned}\right\}\quad(2.4.41)$$

<u>Step 3</u> : <u>Estimate of</u> $\hat{E}(\hat{b}(\hat{p}-\hat{\pi}_{k_1}\hat{p})(\hat{q}-\hat{\pi}_{\ell_1}\hat{q}))$, $\forall\hat{b}\in W^{m+1,\infty}(\hat{K})$, $\forall\hat{p}\in P_{n_\alpha-|\nu|}$, $\forall\hat{q}\in P_{n_\beta-|\mu|}$.

First, we get

$$|\hat{E}(\hat{b}(\hat{p}-\hat{\pi}_{k_1}\hat{p})(\hat{q}-\hat{\pi}_{\ell_1}\hat{q}))| \leq C|\hat{b}|_{0,\infty,\hat{K}}|\hat{p}-\hat{\pi}_{k_1}\hat{p}|_{0,\hat{K}}|\hat{q}-\hat{\pi}_{\ell_1}\hat{q}|_{0,\hat{K}} .$$

Since the hypothesis $k+\ell-|\nu|-|\mu| \geq m+1$ and (2.4.31) imply $k_1+\ell_1 \geq m+1$, we can arbitrarily choose the integers i and j with $0 \leq i \leq k_1$, $0 \leq j \leq \ell_1$ *and* $i+j = m+1$ so that inequalities of type (2.4.37) give

$$\left.\begin{aligned} &|\hat{E}(\hat{b}(\hat{p}-\hat{\pi}_{k_1}\hat{p})(\hat{q}-\hat{\pi}_{\ell_1}\hat{q}))| \leq C|\hat{b}|_{0,\infty,\hat{K}}|\hat{p}|_{i,\hat{K}}|\hat{q}|_{j,\hat{K}} , \\ &\forall\hat{b} \in W^{m+1,\infty}(\hat{K}), \quad \forall\hat{p} \in P_{n_\alpha-|\nu|} , \forall\hat{q} \in P_{n_\beta-|\mu|} . \end{aligned}\right\} \qquad (2.4.42)$$

<u>Step 4</u> : <u>Derivation of the estimate</u> (2.4.26)

When $k+\ell-|\nu|-|\mu| \geq m+1$, to obtain the final estimate (2.4.26), we use the equalities (2.4.28)(2.4.30) and, in the inequalities (2.4.33) (2.4.38) to (2.4.42), we substitute the following inequalities (see CIARLET [1978, § 3.1]) :

$$\left.\begin{aligned} &|\hat{b}|_{i,\infty,\hat{K}} \leq Ch_K^i|b|_{i,\infty,K} , \quad i = 0,\dots,m+1 , \\ &|\hat{p}|_{i,\hat{K}} \leq Ch_K^i|\det(B_K)|^{-1/2}|p|_{i,K} , \quad i = 0,\dots,k_1 , \\ &|\hat{q}|_{j,\hat{K}} \leq Ch_K^j|\det(B_K)|^{-1/2}|q|_{j,K} , \quad j = 0,\dots,\ell_1 . \end{aligned}\right\} \qquad (2.4.43)$$

Then, it remains to replace p and q by $\partial^\nu v$ and $\partial^\mu w$, respectively.

(ii) <u>If</u> $|\nu| = \alpha = k$, $|\mu| = \beta = \ell$ <u>and</u> $m=0$ $(\alpha,\beta = 1,2)$

We want to prove that

$$\left.\begin{aligned} &|E_K(b\partial^\nu v\partial^\mu w)| \leq Ch_K\|b\|_{1,\infty,K}\|v\|_{k,K}\|w\|_{\ell,K} , \quad \forall b \in W^{1,\infty}(K), \\ &\forall v \in P_{K\alpha}, \forall w \in P_{K\beta} . \end{aligned}\right\} \qquad (2.4.44)$$

As in the case (i), we set $\partial^\nu v = p$, $\partial^\mu w = q$ and the relation (2.4.28) reduces the study of $E_K(bpq)$ to that of $\hat{E}(\hat{b}\hat{p}\hat{q})$. We get

$$|\hat{E}(\hat{b}\hat{p}\hat{q})| \leq C|\hat{b}|_{0,\infty,\hat{K}}|\hat{p}|_{0,\infty,\hat{K}}|\hat{q}|_{0,\infty,\hat{K}} \leq C\|\hat{b}\|_{1,\infty,\hat{K}}|\hat{p}|_{0,\hat{K}}|\hat{q}|_{0,\hat{K}} .$$

Assumption (2.4.25) and the Lemma 2.4.1 imply

$$|\hat{E}(\hat{b}\hat{p}\hat{q})| \leq C|\hat{b}|_{1,\infty,\hat{K}}|\hat{p}|_{0,\hat{K}}|\hat{q}|_{0,\hat{K}} \quad .$$

Equality (2.4.28) and the estimates (2.4.43) imply

$$\left. \begin{array}{l} |E_K(bpq)| \leq Ch_K|b|_{1,\infty,K}|p|_{0,K}|q|_{0,K} \ , \ \forall b \in W^{1,\infty}(K),\ \forall p \in P_{n_\alpha-|\nu|}, \\ \forall q \in P_{n_\beta-|\mu|} \quad . \end{array} \right\}$$

Then, to obtain (2.4.44) it remains to replace p and q by $\partial^\nu v$ and $\partial^\mu w$, respectively. ∎

Theorem 2.4.4 :

Let us consider two finite element families (K,P_{K1},Σ_{K1}) *and* (K,P_{K2},Σ_{K2}) *for which the triangles* K *are in correspondence with a reference triangle* $\hat{K}$ *through an invertible affine mapping* F_K *of type* (2.4.1) *and such that the spaces* P_{K1} *and* P_{K2} *satisfy the inclusions* (2.4.20)(2.4.21). *Let* $m \geq 0$ *be an integer and let* $q \geq 1$, $q > \frac{2}{m+1}$ *any real number. We assume that the numerical integration scheme on the reference triangle* $\hat{K}$ *satisfies the following properties (the parameter* α *takes value* 1 *or* 2 *according to whether* $v \in P_{K\alpha} = P_{K1}$ *or* P_{K2}) *:*

$$\forall\hat{\phi} \in P_{m+n_\beta-\ell}(\hat{K}) \quad , \ \hat{E}(\hat{\phi}) = 0 \text{ if } \ell \leq m \ , \tag{2.4.45}$$

$$\forall\hat{\phi} \in P_m(\hat{K}) \quad , \ \hat{E}(\hat{\phi}) = 0. \tag{2.4.46}$$

Then, there exists a constant $C > 0$, *independent of* K, *such that for all function* $\phi \in W^{m+1,q}(K)$, *for all* $w \in P_{K\beta}$, $\beta = 1,2$, *we have :*

$$|E_K(\phi w)| \leq Ch_K^{m+1}(\text{meas}(K))^{\frac{1}{2}-\frac{1}{q}}\|\phi\|_{m+1,q,K}\|w\|_{\ell,K} \ , \tag{2.4.47}$$

where ℓ *is an integer* $\geq \beta$ *and* $h_K = \text{diam}\ (K)$.

Proof :

For all $\phi \in W^{m+1,q}(K)$ and for all $w \in P_{K\beta}$ the relations (2.4.13),

(2.4.20) and (2.4.21) lead to

$$E_K(\phi w) = \det(B_K)\ \hat{E}(\hat{\phi}\hat{w}) \quad , \quad \hat{\phi} \in W^{m+1,q}(\hat{K}),\ \hat{w} \in P_{n_\beta}(\hat{K}). \qquad (2.4.48)$$

We can write

$$\hat{E}(\hat{\phi}\hat{w}) = \hat{E}(\hat{\phi}\hat{\pi}_\ell \hat{w}) + \hat{E}(\hat{\phi}(\hat{w}-\hat{\pi}_\ell \hat{w})) \quad , \qquad (2.4.49)$$

where $\hat{\pi}_\ell$ denotes the orthogonal projection operator in the HILBERT space $L_2(\hat{K})$ on the closed subspace $P_\ell(\hat{K})$. Then, the proof is similar to that of Theorem 2.4.3. We shall just outline the main ideas :

<u>Step 1</u> : <u>Estimate of</u> $\hat{E}(\hat{\phi}\hat{\pi}_\ell \hat{w}), \forall\hat{\phi} \in W^{m+1,q}(\hat{K}), \forall\hat{w} \in P_{n_\beta}(\hat{K})$:

The assumption (2.4.46) combined with $q > \frac{2}{m+1}$ leads to

$$|\hat{E}(\hat{\psi})| \leq C|\hat{\psi}|_{m+1,q,\hat{K}} \quad , \forall\hat{\psi} \in W^{m+1,q}(\hat{K}) .$$

Thus,

$$|\hat{E}(\hat{\phi}\hat{\pi}_\ell \hat{w})| \leq C \sum_{j=0}^{\min(\ell,m+1)} |\hat{\phi}|_{m+1-j,q,\hat{K}} |\hat{w}|_{j,\hat{K}} . \qquad (2.4.50)$$

<u>Step 2</u> : <u>Estimate of</u> $\hat{E}(\hat{\phi}(\hat{w}-\hat{\pi}_\ell \hat{w}))$ $,\forall\hat{\phi} \in W^{m+1,q}(\hat{K}), \forall\hat{w} \in P_{n_\beta}(\hat{K})$

<u>If</u> $1 \leq \ell \leq m$, we have the inclusions (SOBOLEV'S Embedding Theorem)

$$W^{m+1,q}(\hat{K}) \hookrightarrow W^{m+1-\ell,r}(\hat{K}) \hookrightarrow \mathcal{C}^\circ(\hat{K})$$

where r is given by

$$\left.\begin{array}{l} \frac{1}{r} = \frac{1}{q} - \frac{\ell}{2} \ \text{ if } \ 1 \leq q < \frac{2}{\ell} \ , \\ r \geq 1 \text{ large enough so that } m+1-\ell-\frac{2}{r} > 0 \ \text{ if } \ q \geq \frac{2}{\ell} \ . \end{array}\right\}$$

Then,

$$|\hat{\phi}|_{m+1-\ell,r,\hat{K}} \leq C \sum_{j=0}^{\ell} |\hat{\phi}|_{m+1-j,q,\hat{K}} \quad , \forall\hat{\phi} \in W^{m+1,q}(\hat{K})$$

and, using (2.4.45) (2.4.46),

$$\left.\begin{aligned}&|\hat{E}(\hat{\phi}(\hat{w}-\hat{\pi}_\ell \hat{w}))| \leq C|\hat{\phi}|_{m+1-\ell,r,\hat{K}}|\hat{w}-\hat{\pi}_\ell \hat{w}|_{0,\hat{K}}\ , \ \forall\hat{\phi}\in W^{m+1-\ell,r}(\hat{K}),\\ &\forall\hat{w}\in P_{n_\beta}(\hat{K})\quad .\end{aligned}\right\}$$

Hence, using (2.4.37)

$$\left.\begin{aligned}&|\hat{E}(\hat{\phi}(\hat{w}-\hat{\pi}_\ell \hat{w}))| \leq C\sum_{j=0}^{\ell}|\hat{\phi}|_{m+1-j,q,\hat{K}}|\hat{w}|_{j,\hat{K}}\ , \ \forall\hat{\phi}\in W^{m+1,q}(\hat{K})\\ &\forall\hat{w}\in P_{n_\beta}(\hat{K})\end{aligned}\right\} \quad 2.4.51)$$

If $m+1 \leq \ell$, we have the inclusion $W^{m+1,q}(\hat{K}) \subsetneq \mathcal{C}^\circ(\hat{K})$.

From it, we derive,

$$\left.\begin{aligned}&|\hat{E}(\hat{\phi}(\hat{w}-\hat{\pi}_\ell \hat{w}))| \leq C\|\hat{\phi}\|_{m+1,q,\hat{K}}|\hat{w}|_{\ell,\hat{K}}\ , \ \forall\hat{\phi}\in W^{m+1,q}(\hat{K}),\\ &\forall\hat{w}\in P_{n_\beta}(\hat{K})\end{aligned}\right\} \quad (2.4.52)$$

Step 3 : Final estimate (2.4.47)

Then, it remains to combine the relation (2.4.48) the inequalities (2.4.50) (2.4.51) when $1 \leq \ell \leq m$, the inequalities (2.4.50) (2.4.52) when $m+1 \leq \ell$, the inequalities (2.4.43) and (CIARLET [1978], § 3.1])

$$|\hat{v}|_{p,q,\hat{K}} \leq Ch_K^p|\det(B_K)|^{-\frac{1}{q}}|v|_{p,q,K}\ , \ \forall\hat{v}\in W^{p,q}(\hat{K}) \ . \qquad ■$$

Now, using Theorem 2.4.3 and an argument due to ZLÁMAL [1974], we can prove the condition (2.4.18) of uniform $\vec{V}_h$- ellipticity :

Condition (2.4.18) of uniform $\vec{V}_h$- ellipticity :

Theorem 2.4.5 :

Let $\mathcal{C}_h$ *be a regular family of triangulations of the domain* Ω *satisfying the properties* (2.1.1) (2.1.2). *Let* (K,P_{K1},Σ_{K1}), (K,P_{K2},Σ_{K2}) *be two affine or almost-affine families of finite elements for which the finite dimensional spaces* P_{K1} *and* P_{K2} *verify the inclusions* (2.4.20) (2.4.21). *Let* V_{h1}, V_{h2} *be the two associated finite element spaces satisfying the conditions* (2.1.3) *to* (2.1.5).

Moreover, let the numerical integration scheme (2.4.7) *on the reference triangle* $\hat{K}$ *satisfies the following properties :*

(i) *the integration nodes* $\hat{b}_\ell \in \bar{\hat{K}}$, $\ell = 1, \ldots, L$; (2.4.53)

(ii) $\forall \hat{\phi} \in P_{-2+2\max(n_1, n_2-1)}(\hat{K})$, $\hat{E}(\hat{\phi}) = 0$. (2.4.54)

Then, for given $\mathbf{A}_{IJ} \in W^{1,\infty}(\Omega)$, $1 \leq I,J \leq 12$, *there exist constants* $\beta > 0$ *and* $h_1 > 0$ *independent of* h, *such that*

$$\beta \|\vec{v}_h\|^2 \leq a_h(\vec{v}_h, \vec{v}_h) \quad , \quad \forall \vec{v}_h \in \vec{V}_h \, , \quad \forall h < h_1 \, , \tag{2.4.55}$$

where the bilinear form $a_h(\cdot,\cdot)$ *is defined by the relation* (2.4.15).

Proof :

For all $\vec{v}_h \in \vec{V}_h$, the inclusion $\vec{V}_h \subset \vec{V}$ (see relation (2.1.7)) allows us to write

$$a_h(\vec{v}_h, \vec{v}_h) = a(\vec{v}_h, \vec{v}_h) + a_h(\vec{v}_h, \vec{v}_h) - a(\vec{v}_h, \vec{v}_h) \, , \tag{2.4.56}$$

where the bilinear form $a(\cdot,\cdot)$ is defined by the relation (1.5.1). According to the relation (1.6.7) and to the inclusion $\vec{V}_h \subset \vec{V}$, there exists a constant $\alpha > 0$, independent of h, such that :

$$\alpha \|\vec{v}_h\|^2 \leq a(\vec{v}_h, \vec{v}_h) \quad , \quad \forall \vec{v}_h \in \vec{V}_h \; . \tag{2.4.57}$$

Now, we prove the existence of a constant $C > 0$, independent of h, such that

$$|a_h(\vec{v}_h, \vec{v}_h) - a(\vec{v}_h, \vec{v}_h)| \leq Ch \|\vec{v}_h\|^2 \, , \; \forall \vec{v}_h \in \vec{V}_h \; . \tag{2.4.58}$$

Figure 2.4.1 shows that the hypotheses of the theorem permit us to apply Theorem 2.4.3 to different types of terms which occur in the second member of the following relation :

$$a(\vec{v}_h,\vec{v}_h) - a_h(\vec{v}_h,\vec{v}_h) = \sum_{K\in\mathcal{T}_h} \sum_{I,J=1}^{12} E_K[\mathbf{A}_{IJ}(\mathbf{V}_h)_I(\mathbf{V}_h)_J] .$$

Thus, there exists a constant C, independent of h, such that

$$|a(\vec{v}_h,\vec{v}_h) - a_h(\vec{v}_h,\vec{v}_h)| \leq \sum_{K\in\mathcal{T}_h} \sum_{I,J=1}^{12} |E_K(\mathbf{A}_{IJ}(\mathbf{V}_h)_I(\mathbf{V}_h)_J|$$

$$\leq C \sum_{K\in\mathcal{T}_h} h_K \left(\sum_{I,J=1}^{12} \|\mathbf{A}_{IJ}\|_{1,\infty,K}\right) \|\vec{v}_h\|^2_{\vec{V}(K)}$$

$$\leq Ch\|\vec{v}_h\|^2 \quad , \forall\vec{v}_h \in \vec{V}_h .$$

Combining the inequalities (2.4.56) to (2.4.58), we get

$$a_h(\vec{v}_h,\vec{v}_h) \geq (\alpha-Ch) \|\vec{v}_h\|^2 \quad , \forall\vec{v}_h \in \vec{V}_h \quad .$$

We then obtain inequality (2.4.55) with $\beta = \frac{\alpha}{2}$, $h_1 = \frac{\alpha}{2C}$. ∎

Remark 2.4.2 : Assumption (2.4.53) is convenient because it assures that all the integration nodes are located in $\bar{\Omega}$.

∎

Asymptotic error estimate theorem :

Now, we are able to evaluate the different terms of the inequality (2.4.19) and, thus, to derive an asymptotic estimate of the error $\|\vec{u}-\vec{u}_h\|$ between the solution $\vec{u}$ of the continuous problem (1.4.5) and the solution $\overrightarrow{u_h}$ of the approximate problem (2.4.14). The next Theorem 2.4.6 gives a general result of an error estimate ; specifically, it specifies criteria to test the choice of the numerical integration schemes in order to obtain the same asymptotic error estimate than the one gets in the case of exact integration (see Theorem 2.4.1).

Theorem 2.4.6 :

Let $\mathcal{T}_h$ *be a regular family of triangulations of the domain* Ω *satisfying the properties* (2.1.1) (2.1.2). *Let* (K,P_{K1},Σ_{K1}), (K,P_{K2},Σ_{K2}) *be two affine or almost-affine families of finite elements and let* V_{h1}, V_{h2} *be two associated finite element spaces satisfying the conditions* (2.1.3) *to* (2.1.5) *respectively. Let there exist integers* m_1,m_2,n_1,n_2 *such*

$(V_h)_I$ \ $(V_h)_J$	$v_h \in P_{K1} (\ell=1,\ \lvert\mu\rvert=0)$	$Dv_h (v_h \in P_{K1}, \ell=\lvert\mu\rvert=1)$	$v_h \in P_{K2} (\ell=2, \lvert\mu\rvert=0)$	$Dv_h (v_h \in P_{K2}, \ell=2, \lvert\mu\rvert=1)$	$D^2 v_h (v_h \in P_{K2}, \ell=\lvert\mu\rvert=2)$
$v_h \in P_{K1}$ $k = 1$ $\lvert\nu\rvert = 0$	$\forall \hat{\phi} \in P_0,\ \hat{E}(\hat{\phi}) = 0$	$\forall \hat{\phi} \in P_{n_1-1},\ \hat{E}(\hat{\phi}) = 0$	$\forall \hat{\phi} \in P_0,\ \hat{E}(\hat{\phi}) = 0$	$\forall \hat{\phi} \in P_0,\ \hat{E}(\hat{\phi}) = 0$	$\forall \hat{\phi} \in P_{n_2-2},\ \hat{E}(\hat{\phi}) = 0$
Dv_h $(v_h \in P_{K1})$ $k = \lvert\nu\rvert = 1$	$\forall \hat{\phi} \in P_{n_1-1},\ \hat{E}(\hat{\phi}) = 0$	$\forall \hat{\phi} \in P_{2n_1-2},\ \hat{E}(\hat{\phi}) = 0$	$\forall \hat{\phi} \in P_{n_1-1},\ \hat{E}(\hat{\phi}) = 0$	$\forall \hat{\phi} \in P_{n_1-1},\ \hat{E}(\hat{\phi}) = 0$	$\forall \hat{\phi} \in P_{n_1+n_2-3},\ \hat{E}(\hat{\phi})=0$
$v_h \in P_{K2}$ $k = 2$ $\lvert\nu\rvert = 0$	$\forall \hat{\phi} \in P_0,\ \hat{E}(\hat{\phi}) = 0$	$\forall \hat{\phi} \in P_{n_1-1},\ \hat{E}(\hat{\phi}) = 0$	$\forall \hat{\phi} \in P_0,\ \hat{E}(\hat{\phi}) = 0$	$\forall \hat{\phi} \in P_0,\ \hat{E}(\hat{\phi}) = 0$	$\forall \hat{\phi} \in P_{n_2-2},\ \hat{E}(\hat{\phi}) = 0$
Dv_h $(v_h \in P_{K2})$ $k = 2$ $\lvert\nu\rvert = 1$	$\forall \hat{\phi} \in P_0,\ \hat{E}(\hat{\phi}) = 0$	$\forall \hat{\phi} \in P_{n_1-1},\ \hat{E}(\hat{\phi}) = 0$	$\forall \hat{\phi} \in P_0,\ \hat{E}(\hat{\phi}) = 0$	$\forall \hat{\phi} \in P_0,\ \hat{E}(\hat{\phi}) = 0$	$\forall \hat{\phi} \in P_{n_2-2},\ \hat{E}(\hat{\phi}) = 0$
$D^2 v_h$ $(v_h \in P_{K2})$ $k = \lvert\nu\rvert = 2$	$\forall \hat{\phi} \in P_{n_2-2},\ \hat{E}(\hat{\phi}) = 0$	$\forall \hat{\phi} \in P_{n_1+n_2-3},\ \hat{E}(\hat{\phi}) = 0$	$\forall \hat{\phi} \in P_{n_2-2},\ \hat{E}(\hat{\phi}) = 0$	$\forall \hat{\phi} \in P_{n_2-2},\ \hat{E}(\hat{\phi}) = 0$	$\forall \hat{\phi} \in P_{2n_2-4},\ \hat{E}(\hat{\phi}) = 0$

Figure 2.4.1 : *Estimate of the terms* $E_K[\mathbf{A}_{IJ}(\mathbf{V}_h)_I(\mathbf{V}_h)_J]$, $I,J = 1,\ldots,12$, *using the theorem* 2.4.3 (in each case, we indicate the values of the parameters $k, \ell, \lvert\nu\rvert, \lvert\mu\rvert$, defined in the statement of Theorem 2.4.3, as well as the criterion on the integration scheme ; note that $m = 0$)

that, for all $K \in \mathcal{T}_h$, *we have the inclusions*

$$P_{m_1}(K) \subset P_{K1} \subset P_{n_1}(K) \quad , \quad 1 \le m_1 \le n_1 \quad , \tag{2.4.59}$$

$$P_{m_2}(K) \subset P_{K2} \subset P_{n_2}(K) \quad , \quad 2 \le m_2 \le n_2 \quad , \tag{2.4.60}$$

$$H^{m+\alpha+1}(K) \subsetneq \mathcal{C}^{s_\alpha}(K) \; , \; \alpha = 1,2, \; m = -1+\min(m_1,m_2-1) \; , \tag{2.4.61}$$

where s_α *denotes the greatest order of partial derivatives occuring in the definition of the set of degrees of freedom* $\Sigma_{K\alpha}$, $\alpha = 1,2$.

Moreover, let the integration scheme (2.4.7) *on the reference triangle* $\hat{K}$ *satisfies the following conditions*

(i) *the integration nodes* $\hat{b}_\ell \in \bar{\hat{K}}$, $\ell=1,\ldots,L$; (2.4.62)

(ii) $\forall \hat{\phi} \in P_{-2+2\max(n_1,n_2-1)}(\hat{K})$, $\hat{E}(\hat{\phi}) = 0$. (2.4.63)

Then, if the solution $\vec{u} \in \vec{V}$ *of the continuous problem* (1.4.5) *belongs to the space* $(H^{m+2}(\Omega))^2 \times H^{m+3}(\Omega)$, *if* $\mathbf{A}_{IJ} \in W^{m+1,\infty}(\Omega)$, $1 \le I,J \le 12$, *and if* $p^i \in W^{m+1,q}(\Omega)$, $i=1,2,3$, *for some number* $q > \frac{2}{m+1}$ *with* $q \ge 2$, *there exist constants* $C > 0$ *and* $h_1 > 0$, *independent of* h, *such that, for all* $h < h_1$

$$\|\vec{u}-\vec{u}_h\| \le Ch^{\min(m_1,m_2-1)} \left\{ \left(\sum_{\alpha=1}^{2} \|u_\alpha\|^2_{m+2,\Omega} + \|u_3\|^2_{m+3,\Omega} \right)^{1/2} + \left(\sum_{i=1}^{3} \|p^i\|^q_{m+1,q,\Omega} \right)^{1/q} \right\} , \tag{2.4.64}$$

where $\vec{u}_h \in \vec{V}_h$ *is the solution of the discrete problem.*

<u>Proof</u> :

The conditions to apply Theorem 2.4.5 are satisfied. Hence, the condition (2.4.18) of uniform $\vec{V}_h$- ellipticity is verified and it is possible to apply Theorem 2.4.2. Therefore, we are going to evaluate both terms of the second member of the inequality (2.4.19) observing that a bound for the first term is given for $\overrightarrow{v_h} = \overrightarrow{\pi_h u}$. The proof needs the following four steps :

<u>Step 1</u> : <u>Estimate of</u> $\|\vec{u}-\overrightarrow{\pi_h u}\|$

Results of approximation theory (also used in the proof of Theorem 2.4.1) gives

$$\|\vec{u}-\overrightarrow{\pi_h u}\| \le Ch^{\min(m_1,m_2-1)}\left(\sum_{\alpha=1}^{2} |u_\alpha|^2_{m+2,\Omega}+|u_3|^2_{m+3,\Omega}\right)^{1/2} \qquad (2.4.65)$$

where, on the one hand, $\overrightarrow{\pi_h u} = (\pi_{h1}u_1, \pi_{h1}u_2, \pi_{h2}u_3)$ denotes the $\vec{V}_h$-interpolant of the function $\vec{u}$, on the other hand, C denotes, here and in the following, a constant independent of h.

<u>Step 2</u> : <u>Estimate of</u> $\sup\limits_{\vec{w}_h\in\vec{V}_h} \dfrac{|a(\overrightarrow{\pi_h u},\vec{w}_h)-a_h(\overrightarrow{\pi_h u},\vec{w}_h)|}{\|\vec{w}_h\|}$.

The relations (1.5.1) (2.4.15) and (2.4.12) imply, for all $\vec{w}_h \in \vec{V}_h$,

$$|a(\overrightarrow{\pi_h u},\vec{w}_h) - a_h(\overrightarrow{\pi_h u},\vec{w}_h)| \le \sum_{K\in\mathcal{C}_h} \sum_{I,J=1}^{12} |E_K\{\mathbf{A}_{IJ}(\pi_h\mathbf{U})_I(\mathbf{W}_h)_J\}| \qquad (2.4.66)$$

where, by analogy with (1.5.2), we denote $(\pi_h\mathbf{U})_I$ (resp.$(\mathbf{W}_h)_J$) the Ith (resp. J^{th}) element of the column matrix $\pi_h\mathbf{U}$ (resp. $\mathbf{W}_h$) :

$$^t\pi_h\mathbf{U} = [\pi_{h1}u_1 \; (\pi_{h1}u_1)_{,1} \; (\pi_{h1}u_1)_{,2} \; \pi_{h1}u_2 \; (\pi_{h1}u_2)_{,1} \; (\pi_{h1}u_2)_{,2} \; \pi_{h2}u_3$$
$$(\pi_{h2}u_3)_{,1} \; (\pi_{h2}u_3)_{,2} \; (\pi_{h2}u_3)_{,11} \; (\pi_{h2}u_3)_{,12} \; (\pi_{h2}u_3)_{,22}]$$

(resp.

$$^t\mathbf{W}_h = [w_{1h} \; w_{1h,1} \; w_{1h,2} \; w_{2h} \; w_{2h,1} \; w_{2h,2} \; w_{3h} \; w_{3h,1} \; w_{3h,2} \; w_{3h,11}$$
$$w_{3h,12} \; w_{3h,22}]).$$

Then, Figure 2.4.2 shows that a suitable application of Theorem 2.4.3 to the different kinds of terms which occur at the second member of the inequality (2.4.66) gives

$$\sum_{I,J=1}^{12} |E_K\{\mathbf{A}_{IJ}(\pi_h\mathbf{U})_I(\mathbf{W}_h)_J\}| \le Ch_K^{\min(m_1,m_2-1)}\left(\sum_{I,J=1}^{12} \|\mathbf{A}_{IJ}\|_{\min(m_1,m_2-1),\infty,K}\right)$$
$$\times\left(\|\pi_{h1}u_1\|^2_{H^{m+2}(K)} + \|\pi_{h1}u_2\|^2_{H^{m+2}(K)} + \|\pi_{h2}u_3\|^2_{H^{m+3}(K)}\right)^{1/2} \|\vec{w}_h\|_{\vec{V}(K)} \; .$$

$\pi_h \mathbf{U}$ \ $\mathbf{W}_h$	$w_h \in P_{K1} (\ell=1, \lvert\mu\rvert=0)$	$Dw_h (w_h \in P_{K1}, \ell=\lvert\mu\rvert=1)$	$w_h \in P_{K2} (\ell=2, \lvert\mu\rvert=0)$	$Dw_h (w_h \in P_{K2}, \ell=2, \lvert\mu\rvert=1)$	$D^2 w_h (w_h \in P_{K2}, \ell=\lvert\mu\rvert=2)$
$\pi_{h1} u \in P_{K1}$ $(k=m+2,$ $\lvert\nu\rvert=0)$	$\forall\hat\phi \in P_{m-1+n_1}, \hat E(\hat\phi)=0$ if $1 \le m$; $\forall\hat\phi \in P_m, \hat E(\hat\phi)=0$	$\forall\hat\phi \in P_{m-1+n_1}, \hat E(\hat\phi)=0$ $\forall\hat\phi \in P_m, \hat E(\hat\phi)=0$	$\forall\hat\phi \in P_{m-2+n_2}, \hat E(\hat\phi)=0$ if $2 \le m$; $\forall\hat\phi \in P_m, \hat E(\hat\phi)=0$	$\forall\hat\phi \in P_{m-2+n_2}, \hat E(\hat\phi)=0$ if $1 \le m$; $\forall\hat\phi \in P_m, \hat E(\hat\phi)=0$	$\forall\hat\phi \in P_{m-2+n_2}, \hat E(\hat\phi)=0$ $\forall\hat\phi \in P_m, \hat E(\hat\phi)=0$
$D\pi_{h1} u$ $(\pi_{h1} u \in P_{K1})$ $k=m+2$ $\lvert\nu\rvert=1$	$\forall\hat\phi \in P_{m-1+n_1}, \hat E(\hat\phi)=0$ if $1 \le m$; $\forall\hat\phi \in P_m, \hat E(\hat\phi)=0$	$\forall\hat\phi \in P_{m-1+n_1}, \hat E(\hat\phi)=0$ $\forall\hat\phi \in P_m, \hat E(\hat\phi)=0$	$\forall\hat\phi \in P_{m-2+n_2}, \hat E(\hat\phi)=0$ if $2 \le m$; $\forall\hat\phi \in P_m, \hat E(\hat\phi)=0$	$\forall\hat\phi \in P_{m-2+n_2}, \hat E(\hat\phi)=0$ if $1 \le m$; $\forall\hat\phi \in P_m, \hat E(\hat\phi)=0$	$\forall\hat\phi \in P_{m-2+n_2}, \hat E(\hat\phi)=0$ $\forall\hat\phi \in P_m, \hat E(\hat\phi)=0$
$\pi_{h2} u \in P_{K2}$ $k=m+3$ $\lvert\nu\rvert=0$	$\forall\hat\phi \in P_{m-1+n_1}, \hat E(\hat\phi)=0$ if $1 \le m$; $\forall\hat\phi \in P_m, \hat E(\hat\phi)=0$	$\forall\hat\phi \in P_{m-1+n_1}, \hat E(\hat\phi)=0$ $\forall\hat\phi \in P_m, \hat E(\hat\phi)=0$	$\forall\hat\phi \in P_{m-2+n_2}, \hat E(\hat\phi)=0$ if $2 \le m$; $\forall\hat\phi \in P_m, \hat E(\hat\phi)=0$	$\forall\hat\phi \in P_{m-2+n_2}, \hat E(\hat\phi)=0$ if $1 \le m$; $\forall\hat\phi \in P_m, \hat E(\hat\phi)=0$	$\forall\hat\phi \in P_{m-2+n_2}, \hat E(\hat\phi)=0$ $\forall\hat\phi \in P_m, \hat E(\hat\phi)=0$
$D\pi_{h2} u$ $(\pi_{h2} u \in P_{K2})$ $k=m+3$ $\lvert\nu\rvert=1$	$\forall\hat\phi \in P_{m-1+n_1}, \hat E(\hat\phi)=0$ if $1 \le m$; $\forall\hat\phi \in P_m, \hat E(\hat\phi)=0$	$\forall\hat\phi \in P_{m-1+n_1}, \hat E(\hat\phi)=0$ $\forall\hat\phi \in P_m, \hat E(\hat\phi)=0$	$\forall\hat\phi \in P_{m-2+n_2}, \hat E(\hat\phi)=0$ if $2 \le m$; $\forall\hat\phi \in P_m, \hat E(\hat\phi)=0$	$\forall\hat\phi \in P_{m-2+n_2}, \hat E(\hat\phi)=0$ if $1 \le m$; $\forall\hat\phi \in P_m, \hat E(\hat\phi)=0$	$\forall\hat\phi \in P_{m-2+n_2}, \hat E(\hat\phi)=0$ $\forall\hat\phi \in P_m, \hat E(\hat\phi)=0$
$D^2\pi_{h2} u$ $(\pi_{h2} u \in P_{K2})$ $k=m+3$ $\lvert\nu\rvert=2$	$\forall\hat\phi \in P_{m-1+n_1}, \hat E(\hat\phi)=0$ if $1 \le m$; $\forall\hat\phi \in P_m, \hat E(\hat\phi)=0$	$\forall\hat\phi \in P_{m-1+n_1}, \hat E(\hat\phi)=0$ $\forall\hat\phi \in P_m, \hat E(\hat\phi)=0$	$\forall\hat\phi \in P_{m-2+n_2}, \hat E(\hat\phi)=0$ if $2 \le m$; $\forall\hat\phi \in P_m, \hat E(\hat\phi)=0$	$\forall\hat\phi \in P_{m-2+n_2}, \hat E(\hat\phi)=0$ if $1 \le m$; $\forall\hat\phi \in P_m, \hat E(\hat\phi)=0$	$\forall\hat\phi \in P_{m-2+n_2}, \hat E(\hat\phi)=0$ $\forall\hat\phi \in P_m, \hat E(\hat\phi)=0$

Figure 2.4.2 : Estimate of the terms $E_K\{\mathbf{A}_{IJ}(\pi_h \mathbf{U})_I (\mathbf{W}_h)_J\}$ using Theorem 2.4.3 $(m = -1 + \min(m_1, m_2 - 1))$

Since the P_{K1}- interpolant operator (resp. P_{K2}- interpolant) leaves the space $P_{m+1}(K)$ (resp. $P_{m+2}(K)$) invariant, we get

$$\|\pi_{h1}u_\alpha\|_{m+2,K} \leq \|u_\alpha\|_{m+2,K} + \|u_\alpha - \pi_{h1}u_\alpha\|_{m+2,K} \leq C\|u_\alpha\|_{m+2,K} \ ,\alpha = 1,2,$$

$$(\text{resp. } \|\pi_{h2}u_3\|_{m+3,K} \leq C\|u_3\|_{m+3,K}) \ .$$

Finally, with the hypothesis (2.1.2) and the CAUCHY-SCHWARZ inequality, we have

$$\left.\begin{aligned} &\sup_{\vec{w}_h\in\vec{V}_h} \frac{|a(\overrightarrow{\pi_h u},\vec{w}_h)-a_h(\overrightarrow{\pi_h u},\vec{w}_h)|}{\|\vec{w}_h\|} \leq \\ &\leq Ch^{m+1}\Big(\sum_{I,J=1}^{12} \|\mathbf{A}_{IJ}\|_{m+1,\infty,\Omega}\Big)\Big(\sum_{\alpha=1}^{2} \|u_\alpha\|^2_{m+2,\Omega}+\|u_3\|^2_{m+3,\Omega}\Big)^{1/2} \end{aligned}\right\} \quad (2.4.67)$$

<u>Step 3</u> : <u>Estimate of</u> $\sup_{\vec{w}_h\in\vec{V}_h} \dfrac{|f(\vec{w}_h)-f_h(\vec{w}_h)|}{\|\vec{w}_h\|}$

The relations (1.5.19), (1.5.20), (2.4.16) and (2.4.12) give

$$|f(\vec{w}_h)-f_h(\vec{w}_h)| \leq \sum_{K\in\mathcal{T}_h} |E_K({}^t\mathbf{FW}_h)| \ ,$$

with ${}^t\mathbf{FW}_h = (p^1 w_{1h} + p^2 w_{2h} + p^3 w_{3h})\sqrt{a}$. Figure 2.4.3 shows that a suitable application of the Theorem 2.4.4 to both of these types of terms leads to

$$|f(\vec{w}_h)-f_h(\vec{w}_h)| \leq \sum_{K\in\mathcal{T}_h} |E_K({}^t\mathbf{FW}_h)|$$

$$\leq C \sum_{K\in\mathcal{T}_h} h_K^{m+1}(\text{meas}(K))^{\frac{1}{2}-\frac{1}{q}} \Big(\sum_{i=1}^{3}\|p^i\|^q_{m+1,q,K}\Big)^{\frac{1}{q}} \|\vec{w}_h\|_{\vec{V}(K)}$$

$$\leq Ch^{m+1}(\text{meas}(\Omega))^{\frac{1}{2}-\frac{1}{q}} \Big(\sum_{i=1}^{3} \|p^i\|^q_{m+1,q,\Omega}\Big)^{\frac{1}{q}} \|\vec{w}_h\| \ .$$

To get the last inequality, we use the inequality

$$\sum_K |a_K b_K c_K| \leq \Big(\sum_K |a_K|^\alpha\Big)^{\frac{1}{\alpha}} \Big(\sum_K |b_K|^\beta\Big)^{\frac{1}{\beta}} \Big(\sum_K |c_K|^\gamma\Big)^{\frac{1}{\gamma}} \ ,$$

available for all the real numbers $\alpha \geq 1$, $\beta \geq 1$, $\gamma \geq 1$ satisfying $\frac{1}{\alpha} + \frac{1}{\beta} + \frac{1}{\gamma} = 1$. Here, we take $\frac{1}{\alpha} = \frac{1}{2} - \frac{1}{q}$, $\beta = q$, $\gamma = 2$, and thus, we get the condition $q \geq 2$. Hence,

$$\sup_{\vec{w}_h \in \vec{V}_h} \frac{|f(\vec{w}_h) - f_h(\vec{w}_h)|}{\|\vec{w}_h\|} \leq Ch^{m+1} (\text{meas}(\Omega))^{\frac{1}{2} - \frac{1}{q}} \left(\sum_{i=1}^{3} \|p^i\|^q_{m+1,q,\Omega} \right)^{\frac{1}{q}} \quad (2.4.68)$$

1) $E_K(p^\alpha w_{\alpha h}\sqrt{a})$, $p^\alpha\sqrt{a} \in W^{m+1,q}(K)$, $w_{\alpha h} \in P_{K1}$, $\alpha = 1,2$:

Then $\ell = 1$ and we need

$$\left.\begin{array}{l} \forall\hat{\phi} \in P_{m+n_1-1}(\hat{K}),\ \hat{E}(\hat{\phi}) = 0, \text{ if } 1 \leq m, \\ \forall\hat{\phi} \in P_m(\hat{K}),\ \hat{E}(\hat{\phi}) = 0. \end{array}\right\}$$

2) $E_K(p^3 w_{3h}\sqrt{a})$, $p^3\sqrt{a} \in W^{m+1,q}(K)$, $w_{3h} \in P_{K2}$:

Then $\ell = 2$ and we need

$$\left.\begin{array}{l} \forall\hat{\phi} \in P_{m+n_2-2}(\hat{K}),\ \hat{E}(\hat{\phi}) = 0, \text{ if } 2 \leq m, \\ \forall\hat{\phi} \in P_m(\hat{K}),\ \hat{E}(\hat{\phi}) = 0. \end{array}\right\}$$

Figure 2.4.3 : Estimate of the term $E_K({}^t\mathbf{F}\mathbf{V})$ using Theorem 2.4.4

$(m = -1 + \min(m_1, m_2 - 1))$

Step 4 : Final estimate (2.4.64)

We are able to apply Theorem 2.4.2. We get the estimate (2.4.64) by substitution of the inequalities (2.4.65) (2.4.67) (2.4.68) in the abstract error estimate (2.4.19) written for $\vec{v}_h = \overrightarrow{\pi_h u}$.

∎

In the case of *composite finite elements* (HSIEH-CLOUGH-TOCHER, complete or reduced) *the criteria on the numerical integration schemes have to be satisfied on each subtriangle* K_i *constituting the element* :

This is *necessary* because in step 2 of the proof of Theorem 2.4.6 we need $\pi_{h1}u_\alpha \in H^{m+2}(K)$ and $\pi_{h2}u_3 \in H^{m+3}(K)$. Generally, these inclusions are not satisfied for composite finite elements, since we have only $P_K \subset H^2(K)$. Thus, we have to work on each subtriangle K_i, $i = 1,2,3$, to get the suitable inclusions.

This is *sufficient*, because to derive the property of $\vec{V}_h$- ellipticity (2.4.18) and to get some estimates on the consistency terms of the inequality (2.4.19) we use the notion of reference set, excluding the use of the notion of reference finite element. Thus, these results are particularly available for the polynomial functions defined on each subtriangle K_i constituting the composite element K.

As noted early, this property is not too inconvenient since, in applications, the integrals on a composite element $K = \bigcup_i K_i$ have to be calculated as a sum of integrals on each of the subtriangles K_i as the interpolant functions change on each subtriangle.

Remark 2.4.3 : When the middle surface of the shell is *plane* and when the loads are *normal* to this plane, i.e., $p^1 = p^2 = 0$, the KOITER model gives the usual equation of the deflection of a *thin elastic plate*. In this case we only have the unknown u_3. With respect to the approximation, we need only introduce the space V_{h2} so that the condition (2.4.63) becomes

$$\forall \hat{\phi} \in P_{2n_2-4}(\hat{K}),\ \hat{E}(\hat{\phi}) = 0 ,$$

which is the usual condition for plate problems (see BERNADOU-DUCATEL [1978]). ∎

Remark 2.4.4 : In this paper, we have assumed that the boundary Γ of the reference domain Ω is *polygonal*, which is frequently the case in practice. When the boundary Γ is *curved* we have

(i) to define and to analyze curved finite elements of class $\mathcal{C}^1$ and
(ii) to generalize all the results of this paragraph in order to take into account the use of such curved elements.

These two studies are done in BERNADOU [1978]. ∎

CHAPTER 3

IMPLEMENTATION

Orientation :

One of the interesting aspects of the implementation of the finite element method is its *modular character* : the program is subdivided in modules, each module being a set of subprograms with a specific assignment. The main interests of this partition lie in the following two points :

(i) *the ability to solve a given problem by a simple connection* of some specific modules like triangulation, interpolation (based on a given finite element), functionnal (based on a given variational form), elementary stiffness matrix, assemblage, solution of a system, presentation of the results,...

(ii) *the ability to interchange some equivalent modules*, for example, interpolation modules based on different types of $\mathcal{C}^1$- elements.

These principles are to the core of MODULEF Code (see BEGIS-PERRONNET [1980] or PERRONNET [1979]).

In this chapter, we begin by recording prerequisites to the construction of *interpolation modules* using the five finite elements described in the Figures 2.1.1 to 2.1.5. Next, for suitable combinations of these finite elements, we indicate how to evaluate the *energy functional modules* and the *second members modules*.

3.1. Interpolation modules :

We outline briefly prerequisites to the implementation of the triangles of type (1) and (2) described in Figures 2.1.1 and 2.1.2. Next, we record similar results for the ARGYRIS triangle and the complete or reduced HSIEH-CLOUGH-TOCHER triangles, respectively described in Figures

2.1.3 to 2.1.5. For the three last elements, a corresponding analysis is detailed in BERNADOU and BOISSERIE [1978a] and BERNADOU, BOISSERIE, and HASSAN [1980].

By using the following indices, we will distinguish between the five finite elements considered :

$$\left.\begin{array}{l} 1 \ : \text{for triangle of type } \underline{1} \\ 2 \ : \text{for triangle of type } \underline{2} \\ A \ : \text{for the } \underline{A}\text{RGYRIS triangle} \\ C \ : \text{for the } \underline{C}\text{omplete HSIEH-CLOUGH-TOCHER triangle} \\ R \ : \text{for the } \underline{R}\text{educed HSIEH-CLOUGH-TOCHER triangle} \end{array}\right\} \qquad (3.1.1)$$

In all the following discussion, we assume that a fixed orthonormal reference system is given in the Euclidean plane $\mathcal{E}^2$, i.e., $(0,\vec{\varepsilon}_1,\vec{\varepsilon}_2)$. The coordinates of a point P are denoted (x,y), i.e.,

$$\overrightarrow{OP} = x\vec{\varepsilon}_1 + y\vec{\varepsilon}_2 \ . \qquad (3.1.2)$$

In particular, we assume in the following that curvilinear coordinates (ξ^1,ξ^2) (see Figure 1.1.1) are defined so that $\xi^1 = x$ and $\xi^2 = y$.

Then, let us consider a non-degenerate triangle with vertices $a_1(x_1,y_1)$, $a_2(x_2,y_2)$, $a_3(x_3,y_3)$. We have $\overrightarrow{a_1a_2} \times \overrightarrow{a_1a_3} \neq \vec{0}$. The *cartesian coordinates* (x,y) of the point P are given by the expressions

$$\left.\begin{array}{l} x = \lambda_1x_1 + \lambda_2x_2 + \lambda_3x_3 \\ \\ y = \lambda_1y_1 + \lambda_2y_2 + \lambda_3y_3 \end{array}\right\} \qquad (3.1.3)$$

where the *barycentric coordinates* $(\lambda_1,\lambda_2,\lambda_3)$ of the triangle satisfy

$$1 = \lambda_1 + \lambda_2 + \lambda_3. \qquad (3.1.4)$$

Conversely,

$$\left.\begin{aligned}
\lambda_1 &= \frac{(x-x_2)(y_2-y_3) + (x_3-x_2)(y-y_2)}{\Delta} \\
\lambda_2 &= \frac{(x-x_3)(y_3-y_1) + (x_1-x_3)(y-y_3)}{\Delta} \\
\lambda_3 &= \frac{(x-x_1)(y_1-y_2) + (x_2-x_1)(y-y_1)}{\Delta} \\
&\text{where} \\
\Delta &= x_1(y_2-y_3) + x_2(y_3-y_1) + x_3(y_1-y_2) \\
&= (x_1-x_3)(y_2-y_3) + (x_2-x_3)(y_3-y_1) \\
&= (\overrightarrow{a_1a_3} \times \overrightarrow{a_2a_3})\cdot(\overrightarrow{e_1} \times \overrightarrow{e_2}) \neq 0.
\end{aligned}\right\} \quad (3.1.5)$$

<u>Implementation of the triangle of type</u> (1) :

With the notations of Figure 2.1.1, the interpolant $\Pi_{K1}v$ of a function $v \in \mathcal{C}^\circ(K)$ is given by

$$\Pi_{K1}v = [DL_1(v)]_{1\times 3} \quad [\lambda 1]_{3\times 1} \tag{3.1.6}$$

where

$$[DL_1(v)]_{1\times 3} = [v(a_1)\ \ v(a_2)\ \ v(a_3)]$$

and

$${}^t[\lambda 1]_{1\times 3} = [\lambda_1\ \ \lambda_2\ \ \lambda_3]$$

So, we obtain

$$\left.\begin{aligned}
\partial_x\Pi_{K1}v &= [DL_1(v)]_{1\times 3}\,[\partial_x\lambda 1]_{3\times 1} \quad , \\
\partial_y\Pi_{K1}v &= [DL_1(v)]_{1\times 3}\,[\partial_y\lambda 1]_{3\times 1} \quad ,
\end{aligned}\right\} \quad (3.1.7)$$

where, with (3.1.5),

$$\left.\begin{aligned}
[\partial_x\lambda 1] &= \frac{1}{\Delta}\{(y_2-y_3)[\partial_1\lambda 1] + (y_3-y_1)[\partial_2\lambda 1] + (y_1-y_2)[\partial_3\lambda 1]\} \\
[\partial_y\lambda 1] &= \frac{1}{\Delta}\{(x_3-x_2)[\partial_1\lambda 1] + (x_1-x_3)[\partial_2\lambda 1] + (x_2-x_1)[\partial_3\lambda 1]\},
\end{aligned}\right\} \quad (3.1.8)$$

and

and

$$[\partial_i \lambda 1] = \left[\frac{\partial \lambda 1}{\partial \lambda_i}\right] .$$

Implementation of the triangle of type (2) :

With the notations of Figure 2.1.2, the interpolant $\Pi_{K2}v$ of a function $v \in \mathcal{C}^\circ(K)$ is given by

$$\Pi_{K2}v = [DL_2(v)]_{1\times 6} \; [p_2]_{6\times 1}$$

where

$$[DL_2(v)]_{1\times 6} = [v(a_1) \; v(a_2) \; v(a_3) \; v(b_1) \; v(b_2) \; v(b_3)] \quad ,$$

$$[p_2]_{6\times 1} = [A_2]_{6\times 6} \; [\lambda 2]_{6\times 1} \quad , \tag{3.1.9}$$

with

$$[A_2] = \begin{bmatrix} 1 & 0 & 0 & 0 & -1 & -1 \\ 0 & 1 & 0 & -1 & 0 & -1 \\ 0 & 0 & 1 & -1 & -1 & 0 \\ 0 & 0 & 0 & 4 & 0 & 0 \\ 0 & 0 & 0 & 0 & 4 & 0 \\ 0 & 0 & 0 & 0 & 0 & 4 \end{bmatrix} \tag{3.1.10}$$

and

$${}^t[\lambda 2] = [\lambda_1^2 \; \lambda_2^2 \; \lambda_3^2 \; \lambda_2\lambda_3 \; \lambda_3\lambda_1 \; \lambda_1\lambda_2] \quad . \tag{3.1.11}$$

So,

$$\Pi_{K2}v = [DL_2(v)]_{1\times 6} \; [A_2]_{6\times 6} \; [\lambda 2]_{6\times 1} \tag{3.1.12}$$

From this expression, we deduce

$$\left.\begin{aligned} \partial_x \Pi_{K2}v &= [DL_2(v)]_{1\times 6} \; [A_2]_{6\times 6} \; [\partial_x \lambda 2]_{6\times 1} \quad , \\ \partial_y \Pi_{K2}v &= [DL_2(v)]_{1\times 6} \; [A_2]_{6\times 6} \; [\partial_y \lambda 2]_{6\times 1} \quad , \end{aligned}\right\} \tag{3.1.13}$$

where $[\partial_x \lambda 2]_{6\times 1}$ and $[\partial_y \lambda 2]_{6\times 1}$ are defined by analogy with (3.1.8).

Implementation of the ARGYRIS triangle

In comparison with the triangles of type (1) or (2), the triangles of ARGYRIS and of HSIEH-CLOUGH-TOCHER present two new difficulties :

(i) the definition of the degrees of freedom of these triangles is "*local*" in the sense that it uses only the geometry of the triangle : see Figure 2.1.3 to 2.1.5. In order to realize the assembly, it is advisable to define the "*global*" degrees of freedom : this amounts to substitute the fixed directions of the vectors $\vec{\varepsilon}_1, \vec{\varepsilon}_2$ for the directions of the sides of the triangle and, next, to take the unit norm ;

(ii) among the degrees of freedom of these elements, there are normal derivatives at midsides, explicitely for ARGYRIS triangle and complete HCT- triangle and implicitely for reduced HCT-triangle. In general, the affine mapping which associates two non degenerate triangles of any shape, leaves invariant the barycentric coordinates of the associated points. On the other hand, such a transformation changes the angles : thus, the normal at a midside of a triangle is generally changed to an oblique angle at corresponding midside of the associated triangle. Nevertheless, the joint use of barycentric coordinates and *eccentricity parameters* allows to give explicitely the basis functions of these finite elements for triangles of *any shapes*.

With the notations of Figure 3.1.1, the eccentricity parameters are defined by the relations

$$\eta_i = 2\,\frac{\overline{c_i b_i}}{\overline{a_{i-1} a_{i+1}}}\ ,\ i = 1,2,3\ ;\ \eta_i \in \mathbb{R}\ . \qquad (3.1.14)$$

It is easy to evaluate these parameters by using the coordinates of the three vertices of the triangle, i.e.,

$$\eta_i = \frac{(\ell_{i+2})^2 - (\ell_{i+1})^2}{\ell_i^2}\ ,\ i = 1,2,3, \qquad (3.1.15)$$

where

$$\ell_i = \sqrt{(x_{i+1} - x_{i-1})^2 + (y_{i+1} - y_{i-1})^2}\ ,\ i = 1,2,3, \qquad (3.1.16)$$

denotes the length of the side $a_{i-1}a_{i+1}$ of the triangle.

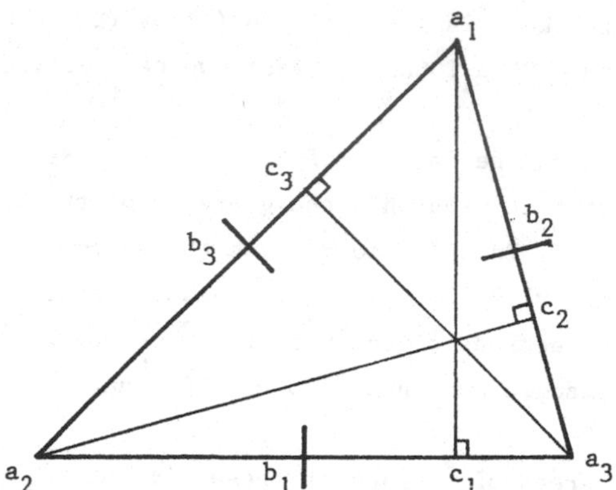

Figure 3.1.1 : Definition of the eccentricity parameters

With the notations of Figure 2.1.3, the interpolant $\Pi_{KA}v$ can be written

$$\Pi_{KA}v = [DLLC_A(v)]_{1\times 21}\ [p_A]_{21\times 1} \tag{3.1.17}$$

where $[DLLC_A(v)]$ denotes the set of *local* degrees of freedom of the function v given as follows

$$\left.\begin{aligned}
[DLLC_A(v)] = \Big[& v(a_1)\ v(a_2)\ v(a_3)\ Dv(a_1)(a_3-a_1)\ Dv(a_1)(a_2-a_1) \\
& Dv(a_2)(a_1-a_2)\ Dv(a_2)(a_3-a_2)\ Dv(a_3)(a_2-a_3)\ Dv(a_3)(a_1-a_3) \\
& D^2v(a_1)(a_3-a_1)^2\ D^2v(a_1)(a_1-a_2)^2\ D^2v(a_2)(a_1-a_2)^2 \\
& D^2v(a_2)(a_2-a_3)^2\ D^2v(a_3)(a_2-a_3)^2\ D^2v(a_3)(a_3-a_1)^2 \\
& D^2v(a_1)(a_2-a_3)^2\ D^2v(a_2)(a_3-a_1)^2\ D^2v(a_3)(a_1-a_2)^2 \\
& Dv(b_1)(a_1-c_1)\ Dv(b_2)(a_2-c_2)\ Dv(b_3)(a_3-c_3)\Big] \quad .
\end{aligned}\right\} \tag{3.1.18}$$

Moreover, the basis polynomials can be written in the following matrix form - see Figure 3.1.2 :

$$[p_A]_{21\times 1} = [A_A]_{21\times 21}\ [\lambda 5]_{21\times 1} \quad . \tag{3.1.19}$$

It is possible to express the *local* degrees of freedom $[DLLC_A(v)]$

$$
\begin{bmatrix} p_1^0 \\ p_2^0 \\ p_3^0 \\ p_{1,3}^1 \\ p_{1,2}^1 \\ p_{2,1}^1 \\ p_{2,3}^1 \\ p_{3,2}^1 \\ p_{3,1}^1 \\ p_{1,2}^2 \\ p_{1,3}^2 \\ p_{2,3}^2 \\ p_{2,1}^2 \\ p_{3,1}^2 \\ p_{3,2}^2 \\ p_{1,1}^2 \\ p_{2,2}^2 \\ p_{3,3}^2 \\ p_{\perp,1}^1 \\ p_{\perp,2}^1 \\ p_{\perp,3}^1 \end{bmatrix}
= \frac{1}{4}
\left[\begin{array}{ccc|cccccc|cccccc|ccc|ccc}
4 & 0 & 0 & 20 & 20 & 0 & 0 & 0 & 0 & 40 & 40 & 0 & 0 & 0 & 0 & 80 & 0 & 0 & 0 & 60(1+\eta_2) & 60(1-\eta_3) \\
 & 4 & 0 & 0 & 0 & 20 & 20 & 0 & 0 & 0 & 0 & 40 & 40 & 0 & 0 & 0 & 80 & 0 & 60(1-\eta_1) & 0 & 60(1+\eta_3) \\
 & & 4 & 0 & 0 & 0 & 0 & 20 & 20 & 0 & 0 & 0 & 0 & 40 & 40 & 0 & 0 & 80 & 60(1+\eta_1) & 60(1-\eta_2) & 0 \\
\hline
 & & & 4 & 0 & 0 & 0 & 0 & 0 & 16 & 0 & 0 & 0 & 0 & 0 & 16 & 0 & 0 & 0 & 34+14\eta_2 & -20 \\
 & & & & 4 & 0 & 0 & 0 & 0 & 0 & 16 & 0 & 0 & 0 & 0 & 16 & 0 & 0 & 0 & -20 & 34-14\eta_3 \\
 & & & & & 4 & 0 & 0 & 0 & 0 & 0 & 16 & 0 & 0 & 0 & 0 & 16 & 0 & -20 & 0 & 34+14\eta_3 \\
 & & & & & & 4 & 0 & 0 & 0 & 0 & 0 & 16 & 0 & 0 & 0 & 16 & 0 & 34-14\eta_1 & 0 & -20 \\
 & & & & & & & 4 & 0 & 0 & 0 & 0 & 0 & 16 & 0 & 0 & 0 & 16 & 34+14\eta_1 & -20 & 0 \\
 & & & & & & & & 4 & 0 & 0 & 0 & 0 & 0 & 16 & 0 & 0 & 16 & -20 & 34-14\eta_2 & 0 \\
\hline
 & & & & & & & & & 2 & 0 & 0 & 0 & 0 & 0 & 2 & 0 & 0 & 0 & 3+\eta_2 & -2 \\
 & & & & & & & & & & 2 & 0 & 0 & 0 & 0 & 2 & 0 & 0 & 0 & -2 & 3-\eta_3 \\
 & & & & & & & & & & & 2 & 0 & 0 & 0 & 0 & 2 & 0 & -2 & 0 & 3+\eta_3 \\
 & & & & & & & & & & & & 2 & 0 & 0 & 0 & 2 & 0 & 3-\eta_1 & 0 & -2 \\
 & & & & & & & & & & & & & 2 & 0 & 0 & 0 & 2 & 3+\eta_1 & -2 & 0 \\
 & & & & & & & & & & & & & & 2 & 0 & 0 & 2 & -2 & 3-\eta_2 & 0 \\
\hline
 & & & & & & & & & & & & & & & -2 & 0 & 0 & 0 & 2 & 2 \\
 & & & & & & & & & & & & & & & & -2 & 0 & 2 & 0 & 2 \\
 & & & & & & & & & & & & & & & & & -2 & 2 & 2 & 0 \\
\hline
 & & & & & & & & & & & & & & & & & & 64 & 0 & 0 \\
 & & & & & & & & & & & & & & & & & & & 64 & 0 \\
 & & & & & & & & & & & & & & & & & & & & 64
\end{array}\right]
\begin{bmatrix} \lambda_1^5 \\ \lambda_2^5 \\ \lambda_3^5 \\ \lambda_1^4\lambda_3 \\ \lambda_1^4\lambda_2 \\ \lambda_2^4\lambda_1 \\ \lambda_2^4\lambda_3 \\ \lambda_3^4\lambda_2 \\ \lambda_3^4\lambda_1 \\ \lambda_1^3\lambda_3^2 \\ \lambda_1^3\lambda_2^2 \\ \lambda_2^3\lambda_1^2 \\ \lambda_2^3\lambda_3^2 \\ \lambda_3^3\lambda_2^2 \\ \lambda_3^3\lambda_1^2 \\ \lambda_1^3\lambda_2\lambda_3 \\ \lambda_1\lambda_2^3\lambda_3 \\ \lambda_1\lambda_2\lambda_3^3 \\ \lambda_1\lambda_2^2\lambda_3^2 \\ \lambda_1^2\lambda_2\lambda_3^2 \\ \lambda_1^2\lambda_2^2\lambda_3 \end{bmatrix}
$$

Figure 3.1.2 : Basis polynomials (shape functions) for ARGYRIS triangle by using the *global* degrees of freedom $[DLGL_A(v)]$, i.e.,

$$\left.\begin{aligned}[DLGL_A(v)] = \Big[& v(a_1)\ v(a_2)\ v(a_3)\ \frac{\partial v}{\partial x}(a_1)\ \frac{\partial v}{\partial y}(a_1)\ \frac{\partial v}{\partial x}(a_2)\ \frac{\partial v}{\partial y}(a_2) \\ & \frac{\partial v}{\partial x}(a_3)\ \frac{\partial v}{\partial y}(a_3)\ \frac{\partial^2 v}{\partial x^2}(a_1)\ \frac{\partial^2 v}{\partial x \partial y}(a_1)\ \frac{\partial^2 v}{\partial y^2}(a_1) \\ & \frac{\partial^2 v}{\partial x^2}(a_2)\ \frac{\partial^2 v}{\partial x \partial y}(a_2)\ \frac{\partial^2 v}{\partial y^2}(a_2)\ \frac{\partial^2 v}{\partial x^2}(a_3)\ \frac{\partial^2 v}{\partial x \partial y}(a_3) \\ & \frac{\partial^2 v}{\partial y^2}(a_3)\ \frac{\partial v}{\partial \nu_1}(b_1)\ \frac{\partial v}{\partial \nu_2}(b_2)\ \frac{\partial v}{\partial \nu_3}(b_3)\Big] \ .\end{aligned}\right\} \quad (3.1.20)$$

By $\frac{\partial}{\partial \nu_i}$ we denote the unit normal derivative to the side $a_{i-1}\ a_{i+1}$ oriented so that $\vec{\nu_i} \cdot \vec{e_x} > 0$, or, $\{\vec{\nu_i} \cdot \vec{e_x} = 1 \text{ if } \vec{\nu_i} \cdot \vec{e_x} = 0\}$. We obtain

$$[DLLC_A(v)]_{1\times 21} = [DLGL_A(v)]_{1\times 21}\ [D_A]_{21\times 21} \qquad (3.1.21)$$

with

$$[D_A]_{21\times 21} \quad \begin{bmatrix} 1 & & & & & & & & & \\ & 1 & & & & & & \bigcirc & & \\ & & 1 & & & & & & & \\ & & & d_1 & & & & & & \\ & & & & d_2 & & & & & \\ & & & & & d_3 & & & & \\ & \bigcirc & & & & & d_4 & & & \\ & & & & & & & n_1 & & \\ & & & & & & & & n_2 & \\ & & & & & & & & & n_3 \end{bmatrix} \qquad (3.1.22)$$

In order to define $[D_A]$, we use the following quantities

$$[d_i]_{2\times 2} = \begin{bmatrix} x_{i-1}-x_i & x_{i+1}-x_i \\ y_{i-1}-y_i & y_{i+1}-y_i \end{bmatrix} \ , \ i = 1,2,3, \qquad (3.1.23)$$

$$[d_4]_{9\times 9} = \left[\begin{array}{ccc|ccc|ccc} X_{31}X_{31} & X_{12}X_{12} & 0 & 0 & 0 & 0 & X_{23}X_{23} & 0 & 0 \\ 2X_{31}Y_{31} & 2X_{12}Y_{12} & 0 & 0 & 0 & 0 & 2X_{23}Y_{23} & 0 & 0 \\ Y_{31}Y_{31} & Y_{12}Y_{12} & 0 & 0 & 0 & 0 & Y_{23}Y_{23} & 0 & 0 \\ \hline 0 & 0 & X_{12}X_{12} & X_{23}X_{23} & 0 & 0 & 0 & X_{31}X_{31} & 0 \\ 0 & 0 & 2X_{12}Y_{12} & 2X_{23}Y_{23} & 0 & 0 & 0 & 2X_{31}Y_{31} & 0 \\ 0 & 0 & Y_{12}Y_{12} & Y_{23}Y_{23} & 0 & 0 & 0 & Y_{31}Y_{31} & 0 \\ \hline 0 & 0 & 0 & 0 & X_{23}X_{23} & X_{31}X_{31} & 0 & 0 & X_{12}X_{12} \\ 0 & 0 & 0 & 0 & 2X_{23}Y_{23} & 2X_{31}Y_{31} & 0 & 0 & 2X_{12}Y_{12} \\ 0 & 0 & 0 & 0 & Y_{23}Y_{23} & Y_{31}Y_{31} & 0 & 0 & Y_{12}Y_{12} \end{array}\right] \tag{3.1.24}$$

where

$$X_{ij} = x_i - x_j,\ Y_{ij} = y_i - y_j,\ 1 \le i,j \le 3,$$

as well as the parameters

$$n_i = \left\{\begin{array}{ll} |\overrightarrow{c_i a_i}|\ \dfrac{XCAI}{|XCAI|} & \text{si } XCAI \neq 0 \\ |\overrightarrow{c_i a_i}|\ \dfrac{YCAI}{|YCAI|} & \text{si } XCAI = 0 \end{array}\right\},\ 1,2,3\ . \tag{3.1.25}$$

The quantities XCAI and YCAI are defined by relations

$$\left.\begin{array}{l} XCAI = x_i - \frac{1}{2}(1-\eta_i)x_{i+1} - \frac{1}{2}(1+\eta_1)x_{i-1} \\ YCAI = y_i - \frac{1}{2}(1-\eta_i)y_{i+1} - \frac{1}{2}(1+\eta_i)y_{i-1} \end{array}\right\} \tag{3.1.26}$$

while

$$|\overrightarrow{c_i a_i}| = [(XCAI)^2 + (YCAI)^2]^{1/2}\ . \tag{3.1.27}$$

Combining relations (3.1.17) (3.1.19) and (3.1.21), we easily obtain

$$\Pi_{KA}v = [DLGL_A(v)]_{1\times 21}\ [D_A]_{21\times 21}\ [A_A]_{21\times 21}\ [\lambda 5]_{21\times 1} \tag{3.1.28}$$

Hence, we have

$$\left.\begin{aligned}
\partial_x \pi_{KA} v &= [DLGL_A(v)][D_A][A_A][\partial_x \lambda 5] \ , \\
\partial_y \pi_{KA} v &= [DLGL_A(v)][D_A][A_A][\partial_y \lambda 5] \ , \\
\partial_{xx} \pi_{KA} v &= [DLGL_A(v)][D_A][A_A][\partial_{xx} \lambda 5] \ , \\
\partial_{xy} \pi_{KA} v &= [DLGL_A(v)][D_A][A_A][\partial_{xy} \lambda 5] \ , \\
\partial_{yy} \pi_{KA} v &= [DLGL_A(v)][D_A][A_A][\partial_{yy} \lambda 5] \ ,
\end{aligned}\right\} \qquad (3.1.29)$$

where the matrices $[\partial_x \lambda 5]$ and $[\partial_y \lambda 5]$ are defined by analogy with the relations (3.1.8). In the same way, we obtain

$$\left.\begin{aligned}
[\partial_{xx}\lambda 5] = \frac{1}{\Delta^2} \{ & (y_2-y_3)^2[\partial_{11}\lambda 5] + (y_3-y_1)^2[\partial_{22}\lambda 5] + (y_1-y_2)^2[\partial_{33}\lambda 5] \\
& + 2(y_2-y_3)(y_3-y_1)[\partial_{12}\lambda 5] \\
& + 2(y_3-y_1)(y_1-y_2)[\partial_{23}\lambda 5] \\
& + 2(y_1-y_2)(y_2-y_3)[\partial_{31}\lambda 5]\} \ , \\
[\partial_{xy}\lambda 5] = \frac{1}{\Delta^2} \{ & (y_2-y_3)(x_3-x_2)[\partial_{11}\lambda 5] + (y_3-y_1)(x_1-x_3)[\partial_{22}\lambda 5] + \\
& + (y_1-y_2)(x_2-x_1)[\partial_{33}\lambda 5] + \\
& + [(y_2-y_3)(x_1-x_3) + (y_3-y_1)(x_3-x_2)][\partial_{12}\lambda 5] \\
& + [(y_3-y_1)(x_2-x_1) + (y_1-y_2)(x_1-x_3)][\partial_{23}\lambda 5] \\
& + [(y_1-y_2)(x_3-x_2) + (y_2-y_3)(x_2-x_1)][\partial_{31}\lambda 5]\} , \\
[\partial_{yy}\lambda 5] = \frac{1}{\Delta^2} \{ & (x_3-x_2)^2[\partial_{11}\lambda 5] + (x_1-x_3)^2[\partial_{22}\lambda 5] + (x_2-x_1)^2[\partial_{33}\lambda 5] \\
& + 2(x_3-x_2)(x_1-x_3)[\partial_{12}\lambda 5] + 2(x_1-x_3)(x_2-x_1)[\partial_{23}\lambda 5] \\
& + 2(x_2-x_1)(x_3-x_2)[\partial_{31}\lambda 5]\} \ ,
\end{aligned}\right\} \qquad (3.1.30)$$

where

$$[\partial_{ii}\lambda 5] = \left[\frac{\partial^2 \lambda 5}{\partial \lambda_i \partial \lambda_j}\right] \ .$$

<u>Remark 3.1.1</u> : <u>The case of degrees of freedom located on the boundary</u>

Let a be a vertex of a triangle located on Γ_0 such that the tangent to Γ_0 at point a is not parallel to coordinate axis. Then, in section 6.3., we shall see that it is preferable to substitute a new set of degrees of freedom (*the border degrees of freedom*) for the usual set of global degrees of freedom defined at point a. The definition of this set of border degrees-of-freedom uses the directions of the tangent and the normal to Γ_0 at point a. If a is a *salient point* of Γ_0 admitting two distinct half tangents, then we shall use these two directions instead of the tangent and the normal. ∎

Implementation of the complete HSIEH-CLOUGH-TOCHER triangle

With the notations of Figure 2.1.4, the interpolant $\Pi_{K_iC}v$ of the function v on the subtriangle K_i of triangle K can be written

$$\Pi_{K_iC}v = [DLLC_{C_i}(v)]_{1\times 12}\,[R_{C_i}]_{12\times 1} \tag{3.1.31}$$

where, on the one hand,

$$\left.\begin{aligned}[DLLC_{C_i}(v)] = [&v(a_i)\; v(a_{i+1})\; v(a_{i+2})\; Dv(a_i)(a_{i+2}-a_i)\\ &Dv(a_i)(a_{i+1}-a_i)\; Dv(a_{i+1})(a_i-a_{i+1})\\ &Dv(a_{i+1})(a_{i+2}-a_{i+1})\; Dv(a_{i+2})(a_{i+1}-a_{i+2})\\ &Dv(a_{i+2})(a_i-a_{i+2})\; Dv(b_i)(a_i-c_i)\\ &Dv(b_{i+1})(a_{i+1}-c_{i+1})\; Dv(b_{i+2})(a_{i+2}-c_{i+2})]\end{aligned}\right\} \tag{3.1.32}$$

and where, on the other hand, the basis polynomials R_{C_i} are given by (see Figure 3.1.3 and BERNADOU-HASSAN [1981])

$$[R_{C_i}]_{12\times 1} = [A_{C_i}]_{12\times 10}\,[\lambda 3_i]_{10\times 1} \quad . \tag{3.1.33}$$

Moreover, we can express the local degrees of freedom $[DLLC_{C_i}(v)]$ using the set of global degrees of freedom $[DLGL_{C_i}(v)]$, i.e.,

$$\left.\begin{aligned}[DLGL_{C_i}(v)] = [&v(a_i)\; v(a_{i+1})\; v(a_{i+2})\; \frac{\partial v}{\partial x}(a_i)\; \frac{\partial v}{\partial y}(a_i)\; \frac{\partial v}{\partial x}(a_{i+1})\\ &\frac{\partial v}{\partial y}(a_{i+1})\; \frac{\partial v}{\partial x}(a_{i+2})\; \frac{\partial v}{\partial y}(a_{i+2})\; \frac{\partial v}{\partial \nu_i}(b_i)\; \frac{\partial v}{\partial \nu_{i+1}}(b_{i+1})\\ &\frac{\partial v}{\partial \nu_{i+2}}(b_{i+2})] \quad .\end{aligned}\right\} \tag{3.1.34}$$

$$\begin{bmatrix} r^{o}_{i,i} \\ r^{o}_{i,i+1} \\ r^{o}_{i,i+2} \\ r^{1}_{i,i,i+2} \\ r^{1}_{i,i,i+1} \\ r^{1}_{i,i+1,i} \\ r^{1}_{i,i+1,i+2} \\ r^{1}_{i,i+2,i+1} \\ r^{1}_{i,i+2,i} \\ r^{\perp}_{i,i} \\ r^{\perp}_{i,i+1} \\ r^{\perp}_{i,i+2} \end{bmatrix} = \begin{bmatrix} -\frac{1}{2}(\eta_{i+1}-\eta_{i+2}) & 0 & 0 & \frac{3}{2}(3+\eta_{i+1}) & \frac{3}{2}(3-\eta_{i+2}) & 0 & 0 & 0 & 0 & 0 \\ \frac{1}{2}(1-2\eta_i-\eta_{i+2}) & 1 & 0 & -\frac{3}{2}(1-\eta_i) & \frac{3}{2}(\eta_i+\eta_{i+2}) & 3 & 3 & 0 & 0 & 3(1-\eta_i) \\ \frac{1}{2}(1+2\eta_i+\eta_{i+1}) & 0 & 1 & -\frac{3}{2}(\eta_i+\eta_{i+1}) & -\frac{3}{2}(1+\eta_i) & 0 & 0 & 3 & 3 & 3(1+\eta_i) \\ -\frac{1}{12}(1+\eta_{i+1}) & 0 & 0 & \frac{1}{4}(7+\eta_{i+1}) & -\frac{1}{2} & 0 & 0 & 0 & 0 & 0 \\ -\frac{1}{12}(1-\eta_{i+2}) & 0 & 0 & -\frac{1}{2} & \frac{1}{4}(7-\eta_{i+2}) & 0 & 0 & 0 & 0 & 0 \\ -\frac{1}{12}(7+\eta_{i+2}) & 0 & 0 & \frac{1}{2} & \frac{1}{4}(5+\eta_{i+2}) & 1 & 0 & 0 & 0 & -1 \\ \frac{1}{6}(4-\eta_i) & 0 & 0 & -\frac{1}{4}(3-\eta_i) & -\frac{1}{4}(5-\eta_i) & 0 & 1 & 0 & 0 & \frac{1}{2}(3-\eta_i) \\ \frac{1}{6}(4+\eta_i) & 0 & 0 & -\frac{1}{4}(5+\eta_i) & -\frac{1}{4}(3+\eta_i) & 0 & 0 & 1 & 0 & \frac{1}{2}(3+\eta_i) \\ -\frac{1}{12}(7-\eta_{i+1}) & 0 & 0 & \frac{1}{4}(5-\eta_{i+1}) & \frac{1}{2} & 0 & 0 & 0 & 1 & -1 \\ \frac{4}{3} & 0 & 0 & -2 & -2 & 0 & 0 & 0 & 0 & 4 \\ -\frac{2}{3} & 0 & 0 & 2 & 0 & 0 & 0 & 0 & 0 & 0 \\ -\frac{2}{3} & 0 & 0 & 0 & 2 & 0 & 0 & 0 & 0 & 0 \end{bmatrix} \begin{bmatrix} \lambda_i^3 \\ \lambda_{i+1}^3 \\ \lambda_{i+2}^3 \\ \lambda_i^2\lambda_{i+2} \\ \lambda_i^2\lambda_{i+1} \\ \lambda_{i+1}^2\lambda_i \\ \lambda_{i+1}^2\lambda_{i+2} \\ \lambda_{i+2}^2\lambda_{i+1} \\ \lambda_{i+2}^2\lambda_i \\ \lambda_i\lambda_{i+1}\lambda_{i+2} \end{bmatrix}$$

Figure 3.1.3 : Basis polynomials (shape functions) associated with the sub-triangle K_i of the complete H.C.T. Triangle (η_1,η_2,η_3 denote the eccentricity parameters of the triangle K).

Thus, we obtain

$$[DLLC_{C_i}(v)]_{1\times 12} = [DLGL_{C_i}(v)]_{1\times 12}\,[D_{C_i}]_{12\times 12} \qquad (3.1.35)$$

where

$$[D_{C_i}]_{12\times 12} = \begin{bmatrix} 1 &&&&&&&& \\ & 1 &&&&&&& \\ && 1 &&&& \bigcirc && \\ &&& d_i &&&&& \\ &&&& d_{i+1} &&&& \\ &&&&& d_{i+2} &&& \\ &&&&&& n_i && \\ && \bigcirc &&&&& n_{i+1} & \\ &&&&&&&& n_{i+2} \end{bmatrix} \qquad (3.1.36)$$

and where the parameters d_i and n_i are those defined by (3.1.23) and (3.1.25), respectively.

By substituting (3.1.33) and (3.1.35) into (3.1.31), we finally obtain

$$\Pi_{K_iC}v = [DLGL_{C_i}(v)]_{1\times 12}\,[D_{C_i}]_{12\times 12}\,[A_{C_i}]_{12\times 10}\,[\lambda 3_i]_{10\times 1} \qquad (3.1.37)$$

Then, in analogy with (3.1.29), we derive

$$\left.\begin{aligned} \partial_x\pi_{K_iC}v &= [DLGL_{C_i}(v)]\,[D_{C_i}]\,[A_{C_i}]\,[\partial_x\lambda 3_i] \;, \\ \partial_y\pi_{K_iC}v &= [DLGL_{C_i}(v)]\,[D_{C_i}]\,[A_{C_i}]\,[\partial_y\lambda 3_i] \;, \\ \partial_{xx}\pi_{K_iC}v &= [DLGL_{C_i}(v)]\,[D_{C_i}]\,[A_{C_i}]\,[\partial_{xx}\lambda 3_i] \;, \\ \partial_{xy}\pi_{K_iC}v &= [DLGL_{C_i}(v)]\,[D_{C_i}]\,[A_{C_i}]\,[\partial_{xy}\lambda 3_i] \;, \\ \partial_{yy}\pi_{K_iC}v &= [DLGL_{C_i}(v)]\,[D_{C_i}]\,[A_{C_i}]\,[\partial_{yy}\lambda 3_i] \;. \end{aligned}\right\} \qquad (3.1.38)$$

<u>Remark 3.1.2</u> : The Remark 3.1.1 is still applicable here. ∎

<u>Implementation of the reduced HSIEH-CLOUGH-TOCHER triangle</u>

With the notations of Figure 2.1.5, the interpolant $\Pi_{K_iR}v$ of the function v on the sub-triangle K_i of triangle K can be written

$$\Pi_{K_i R} v = [DLLC_{R_i}(v)]_{1\times 9} \; [R_{R_i}]_{9\times 1} \tag{3.1.39}$$

where, on the one hand,

$$\left.\begin{aligned}[DLLC_{R_i}(v)] = [&v(a_i) \; v(a_{i+1}) \; v(a_{i+2}) \; Dv(a_i)(a_{i+2}-a_i) \\ &Dv(a_i)(a_{i+1}-a_i) \; Dv(a_{i+1})(a_i-a_{i+1}) \\ &Dv(a_{i+1})(a_{i+2}-a_{i+1}) \; Dv(a_{i+2})(a_{i+1}-a_{i+2}) \\ &Dv(a_{i+2})(a_i-a_{i+2})]\end{aligned}\right\} \tag{3.1.40}$$

and where, on the other hand, the basis polynomials R_{R_i} are given by (see Figure 3.1.4 and BERNADOU-HASSAN [1981])

$$[R_{R_i}]_{9\times 1} = [A_{R_i}]_{9\times 10} \; [\lambda 3_i]_{10\times 1} \tag{3.1.41}$$

Moreover, we can express the local degrees of freedom $[DLLC_{R_i}(v)]$ using the set of global degrees of freedom $[DLGL_{R_i}(v)]$, i.e.,

$$[DLLC_{R_i}(v)]_{1\times 9} = [DLGL_{R_i}(v)]_{1\times 9} \; [D_{R_i}]_{9\times 9} \tag{3.1.42}$$

where

$$\left.\begin{aligned}[DLGL_{R_i}(v)] = [&v(a_i) \; v(a_{i+1}) \; v(a_{i+2}) \; \frac{\partial v}{\partial x}(a_i) \; \frac{\partial v}{\partial y}(a_i) \\ &\frac{\partial v}{\partial x}(a_{i+1}) \; \frac{\partial v}{\partial y}(a_{i+1}) \; \frac{\partial v}{\partial x}(a_{i+2}) \; \frac{\partial v}{\partial y}(a_{i+2})],\end{aligned}\right\} \tag{3.1.43}$$

and

$$[D_{R_i}] = \begin{bmatrix} 1 & & & & & \\ & 1 & & & \bigcirc & \\ & & 1 & & & \\ & & & d_i & & \\ & \bigcirc & & & d_{i+1} & \\ & & & & & d_{i+2} \end{bmatrix} \tag{3.1.44}$$

$$
\begin{bmatrix} r^{o}_{i,i} \\ r^{o}_{i,i+1} \\ r^{o}_{i,i+2} \\ r^{1}_{i,i,i+2} \\ r^{1}_{i,i,i+1} \\ r^{1}_{i,i+1,i} \\ r^{1}_{i,i+1,i+2} \\ r^{1}_{i,i+2,i+1} \\ r^{1}_{i,i+2,i} \end{bmatrix} = \begin{bmatrix}
-\frac{1}{2}(\eta_{i+1}-\eta_{i+2}) & 0 & 0 & \frac{3}{2}(3+\eta_{i+1}) & \frac{3}{2}(3-\eta_{i+2}) & 0 & 0 & 0 & 0 & 0 \\
\frac{1}{2}(1-2\eta_i-\eta_{i+2}) & 1 & 0 & -\frac{3}{2}(1-\eta_i) & \frac{3}{2}(\eta_i+\eta_{i+2}) & 3 & 3 & 0 & 0 & 3(1-\eta_i) \\
\frac{1}{2}(1+2\eta_i+\eta_{i+1}) & 0 & 1 & -\frac{3}{2}(\eta_i+\eta_{i+1}) & -\frac{3}{2}(1+\eta_i) & 0 & 0 & 3 & 3 & 3(1+\eta_i) \\
-\frac{1}{4}(1+\eta_{i+1}) & 0 & 0 & \frac{1}{4}(5+3\eta_{i+1}) & \frac{1}{2} & 0 & 0 & 0 & 0 & 0 \\
-\frac{1}{4}(1-\eta_{i+2}) & 0 & 0 & \frac{1}{2} & \frac{1}{4}(5-3\eta_{i+2}) & 0 & 0 & 0 & 0 & 0 \\
\frac{1}{4}(1-\eta_{i+2}) & 0 & 0 & -\frac{1}{2} & -\frac{1}{4}(1-3\eta_{i+2}) & 1 & 0 & 0 & 0 & 1 \\
-\frac{1}{2}\eta_i & 0 & 0 & -\frac{1}{4}(1-3\eta_i) & \frac{1}{4}(1+3\eta_i) & 0 & 1 & 0 & 0 & \frac{1}{2}(1-3\eta_i) \\
\frac{1}{2}\eta_i & 0 & 0 & \frac{1}{4}(1-3\eta_i) & -\frac{1}{4}(1+3\eta_i) & 0 & 0 & 1 & 0 & \frac{1}{2}(1+3\eta_i) \\
\frac{1}{4}(1+\eta_{i+1}) & 0 & 0 & -\frac{1}{4}(1+3\eta_{i+1}) & -\frac{1}{2} & 0 & 0 & 0 & 1 & 1
\end{bmatrix} \begin{bmatrix} \lambda_i^3 \\ \lambda_{i+1}^3 \\ \lambda_{i+2}^3 \\ \lambda_i^2\lambda_{i+2} \\ \lambda_i^2\lambda_{i+1} \\ \lambda_{i+1}^2\lambda_i \\ \lambda_{i+1}^2\lambda_{i+2} \\ \lambda_{i+2}^2\lambda_{i+1} \\ \lambda_{i+2}^2\lambda_i \\ \lambda_i\lambda_{i+1}\lambda_{i+2} \end{bmatrix}
$$

Figure 3.1.4 : Basis polynomials (shape functions) associated with the sub-triangle K_i of the reduced HCT-triangle (η_1,η_2,η_3 denote the eccentricity parameters of the triangle K).

In the last expression, the matrices $[d_i]$ are given by the relations (3.1.23). Substituting (3.1.41) and (3.1.42) into (3.1.39), we obtain

$$\Pi_{K_iR}v = [DLGL_{R_i}(v)]_{1\times 9}\ [D_{R_i}]_{9\times 9}\ [A_{R_i}]_{9\times 10}\ [\lambda 3_i]_{10\times 1} \tag{3.1.45}$$

Then, similarly to (3.1.29), we get

$$\left.\begin{aligned}
\partial_x\pi_{K_iR}v &= [DLGL_{R_i}(v)][D_{R_i}][A_{R_i}][\partial_x\lambda 3_i] \quad ,\\
\partial_y\pi_{K_iR}v &= [DLGL_{R_i}(v)][D_{R_i}][A_{R_i}][\partial_y\lambda 3_i] \quad ,\\
\partial_{xx}\pi_{K_iR}v &= [DLGL_{R_i}(v)][D_{R_i}][A_{R_i}][\partial_{xx}\lambda 3_i] \quad ,\\
\partial_{xy}\pi_{K_iR}v &= [DLGL_{R_i}(v)][D_{R_i}][A_{R_i}][\partial_{xy}\lambda 3_i] \quad ,\\
\partial_{yy}\pi_{K_iR}v &= [DLGL_{R_i}(v)][D_{R_i}][A_{R_i}][\partial_{yy}\lambda 3_i] \quad .
\end{aligned}\right\} \tag{3.1.46}$$

Remark 3.1.3 : The Remark 3.1.1 is still applicable here. ■

3.2. Energy functional and second member modules when the spaces X_{h1} and X_{h2} are constructed using ARGYRIS triangles :

Following (2.2.3) to (2.2.5), the discrete problem under consideration can be stated :

$$\left.\begin{aligned}
&\text{Find } \vec{u_h} \in \vec{V_h} \text{ such that for any } \vec{v_h} \in \vec{V_h}\\
&a_h(\vec{u_h},\vec{v_h}) = f_h(\vec{v_h}) \ ,
\end{aligned}\right\} \tag{3.2.1}$$

where

$$a_h(\vec{u_h},\vec{v_h}) = \sum_{K\in\mathcal{C}_h} \sum_{\ell=1}^{L} \omega_{\ell,K}\ {}^t\mathbf{U}_h(b_{\ell,K})[\mathbf{A}_{IJ}(b_{\ell,K})]\ \mathbf{V}_h(b_{\ell,K}) \tag{3.2.2}$$

$$f_h(\vec{v_h}) = \sum_{K\in\mathcal{C}_h} \sum_{\ell=1}^{L} \omega_{\ell,K}\ {}^t\mathbf{F}(b_{\ell,K})\mathbf{V}_h(b_{\ell,K}) \ . \tag{3.2.3}$$

So in relation (3.1.20), we denote

$DLGL_A(u_1)$ the matrix of global degrees of freedom associated with u_1 ,
$DLGL_A(u_2)$ the matrix of global degrees of freedom associated with u_2 ,
$DLGL_A(u_3)$ the matrix of global degrees of freedom associated with u_3 .

Then, the matrix ${}^t\mathbf{U}_h$, i.e.,

$$
{}^t\mathbf{U}_h = [u_{1h}\; u_{1h,x}\; u_{1h,y}\; u_{2h}\; u_{2h,x}\; u_{2h,y}\; u_{3h}\; u_{3h,x}\; u_{3h,y}\; u_{3h,xx}\; u_{3h,xy}\; u_{3h,yy}] \tag{3.2.4}
$$

can be written on the triangle K as follows :

$$
{}^t\mathbf{U}_h = [DG(\overrightarrow{u_h})]_{1\times 63} \times \begin{bmatrix} DA & 0 & 0 \\ 0 & DA & 0 \\ 0 & 0 & DA \end{bmatrix}_{63\times 63} \left[\begin{array}{ccc|ccc|cccccc} \lambda5 & \partial_x\lambda5 & \partial_y\lambda5 & & 0 & & & & 0 & & & \\ & 0 & & \lambda5 & \partial_x\lambda5 & \partial_y\lambda5 & & & 0 & & & \\ & 0 & & & 0 & & \lambda5 & \partial_x\lambda5 & \partial_y\lambda5 & \partial_{xx}\lambda5 & \partial_{xy}\lambda5 & \partial_{yy}\lambda5 \end{array}\right]_{63\times 12} \tag{3.2.5}
$$

where

$$
[DG(\overrightarrow{u_h})]_{1\times 63} = [DLGL_A(u_{1h})\; ;\; DLGL_A(u_{2h})\; ;\; DLGL_A(u_{3h})] \tag{3.2.6}
$$

and

$$
[DA]_{21\times 21} = [D_A]_{21\times 21}\; [A_A]_{21\times 21} \tag{3.2.7}
$$

The matrix $[A_A]$ and $[D_A]$ are defined by relations (3.1.19) and (3.1.22). Also, let us observe that in the expression (3.2.4) we use variables (x,y) instead of curvilinear coordinates (ξ^1,ξ^2) used into relation (1.5.2). This is convenient in order to avoid any possible confusion between the indices of (ξ^1,ξ^2) and those of $(\lambda_1,\lambda_2,\lambda_3)$ (see for instance relations (3.1.8)).

Following Figure 2.3.1, we need to use a numerical integration scheme *exact for polynomials of degree* 8 (for instance the scheme given in Figure 2.2.3). In this scheme, defined on a general triangle, the integration nodes b_ℓ, $\ell=1,\ldots,16$ are located by their barycentric coordinates. Let ω_ℓ, $\ell=1,\ldots,16$, be the corresponding weights.

Let us set

$$[DT] = \begin{bmatrix} DA & 0 & 0 \\ 0 & DA & 0 \\ 0 & 0 & DA \end{bmatrix}_{63\times 63} \qquad (3.2.8)$$

and

$$[LAMBD] = \left[\begin{array}{c|c|c} \lambda 5 \;\; \partial_x\lambda 5 \;\; \partial_y\lambda 5 & 0 & 0 \\ 0 & \lambda 5 \;\; \partial_x\lambda 5 \;\; \partial_y\lambda 5 & 0 \\ 0 & 0 & \lambda 5 \;\; \partial_x\lambda 5 \;\; \partial_y\lambda 5 \;\; \partial_{xx}\lambda 5 \;\; \partial_{xy}\lambda 5 \;\; \partial_{yy}\lambda 5 \end{array}\right]_{63\times 12} \qquad (3.2.9)$$

Then, we write relations (3.2.2) and (3.2.3) respectively as follows :

$$\left.\begin{array}{l} a_h(\vec{u_h},\vec{v_h}) = \sum\limits_{K\in\mathcal{T}_h} \{\text{aire}(K)[DG(\vec{u_h})]_{1\times 63}\;[DT]_{63\times 63} \\ \{\sum\limits_{\ell=1}^{16} (\;\omega_\ell[LAMBD]_{63\times 12}\;[\mathbf{A}_{IJ}]_{12\times 12}\;{}^t[LAMBD])\;(b_\ell)\}\;{}^t[DT]\;{}^t[DG(\vec{v_h})]\} \end{array}\right\} (3.2.10)$$

$$\left.\begin{array}{l} f_h(\vec{v_h}) = \sum\limits_{K\in\mathcal{T}_h} \{\text{aire}(K)\{\sum\limits_{\ell=1}^{16} (\omega_\ell\;{}^t[\mathbf{F}]_{1\times 12}\;{}^t[LAMBD]_{12\times 63})(b_\ell)\} \\ \qquad {}^t[DT]\;{}^t[DG(\vec{v_h})]\} \end{array}\right\} (3.2.11)$$

<u>Remark 3.2.1</u> : For simplicity, relations (3.2.10) (3.2.11) do not take into account the possible use of *border degrees of freedom* (see Remark 3.1.1 and section 6.3).

3.3. Energy functional and second member modules when the spaces X_{h1} and X_{h2} are constructed using the complete H.C.T.-triangle :

The contents of this section and of the three following are similar to the contents of section 3.2. Thus, we shall simply give the main results. Successively, on each sub-triangle K_i of the triangle K, we have :

$$ {}^t\mathbf{U}_h = [DG_i(\vec{u_h})]_{1\times 36} \quad [DT_i]_{36\times 30} \quad [LAMBD_i]_{30\times 12} \qquad (3.3.1) $$

where, with notations of relation (3.1.37),

$$ [DG_i(\vec{u_h})]_{1\times 36} = [DLGL_{C_i}(u_{1h}) \ ; \ DLGL_{C_i}(u_{2h}) ; \ DLGL_{C_i}(u_{3h})] \qquad (3.3.2) $$

$$ [DT_i] = \begin{bmatrix} DA_{C_i} & 0 & 0 \\ 0 & DA_{C_i} & 0 \\ 0 & 0 & DA_{C_i} \end{bmatrix}_{36\times 30} \qquad (3.3.3) $$

$$ [DA_{C_i}] = [D_{C_i}]_{12\times 12} \quad [A_{C_i}]_{12\times 10} \qquad (3.3.4) $$

$$ [LAMBD_i]_{30\times 12} = $$

$$ \left[\begin{array}{c|c|c} \lambda 3_i \ \partial_x\lambda 3_i \ \partial_y\lambda 3_i & 0 & 0 \\ 0 & \lambda 3_i \ \partial_x\lambda 3_i \ \partial_y\lambda 3_i & 0 \\ 0 & 0 & \lambda 3_i \ \partial_x\lambda 3_i \ \partial_y\lambda 3_i \ \partial_{xx}\lambda 3_i \ \partial_{xy}\lambda 3_i \ \partial_{yy}\lambda 3_i \end{array}\right] \qquad (3.3.5) $$

Following Figure 2.3.1, we need to use a numerical integration scheme *exact for polynomials of degree 4 on every sub-triangle* K_i *of the triangle* K. For instance, we can use the scheme given in Figure 2.2.2. Then, relations (3.2.2) and (3.2.3) yield to

$$ a_h(\vec{u_h},\vec{v_h}) = \sum_{K\in\mathcal{C}_h} \sum_{i=1}^{3} \left\{ [DG_i(\vec{u_h})]_{1\times 36} \quad [M_i]_{36\times 36} \quad {}^t[DG_i(\vec{v_h})]_{36\times 1} \right\} \qquad (3.3.6) $$

where

$$ \left. \begin{aligned} [M_i] = [DT_i]_{36\times 30} \Big\{ & aire(K_i) \sum_{\ell_i=1}^{6} (\omega_{\ell_i} [LAMBD_i]_{30\times 12} \quad [\mathbf{A}_{IJ}]_{12\times 12} \\ & {}^t[LAMBD_i]) \ (b_{\ell_i}) \Big\} \ {}^t[DT_i]_{30\times 36} \end{aligned} \right\} \qquad (3.3.7) $$

and

$$f_h(\vec{v_h}) = \sum_{K\in\mathscr{C}_h} \sum_{i=1}^{3} \left\{ [B_i]_{1\times 36} \ {}^t[DG_i(\vec{v_h})]_{36\times 1} \right\} \tag{3.3.8}$$

with

$$\left. \begin{array}{l} [B_i]_{1\times 36} = \text{aire}(K_i) \left\{ \sum_{\ell_i=1}^{6} (\omega_{\ell_i} \ {}^t[\mathbf{F}]_{1\times 12} \ {}^t[LAMBD_i]_{12\times 30}) \right. \\ \qquad\qquad (b_{\ell_i}) \Big\} \ {}^t[DT_i]_{30\times 36} \end{array} \right\} \tag{3.3.9}$$

<u>Remark 3.3.1</u> : A similar remark to Remark 3.2.1 applies to the relations (3.3.6) and (3.3.8). ∎

<u>3.4. Energy functional and second member modules when the spaces X_{h1} and X_{h2} are constructed using triangles of type (2) and complete H.C.T. triangles, respectively</u>

In a manner parallel to that used in section 3.2, we obtain, on every sub-triangle K_i of the triangle K,

$${}^t\mathbf{U}_h = [DG_i(\vec{u_h})]_{1\times 24} \ [DT_i]_{24\times 22} \ [LAMBD_i]_{22\times 12} \tag{3.4.1}$$

where, with the notations of relations (3.1.12) and (3.1.37), we have

$$[DG_i(\vec{u_h})]_{1\times 24} = [DL_2(u_{1h}) \ ; \ DL_2(u_{2h}) \ ; \ DLGL_{C_i}(u_{3h})] \tag{3.4.2}$$

$$[DT_i]_{24\times 22} = \begin{bmatrix} A_2 & 0 & 0 \\ 0 & A_2 & 0 \\ 0 & 0 & DA_{C_i} \end{bmatrix} \tag{3.4.3}$$

$$[DA_{C_i}]_{12\times 10} = [D_{C_i}]_{12\times 12} \ [A_{C_i}]_{12\times 10} \tag{3.4.4}$$

$$[LAMBD_i]_{22\times 12} =$$

$$\left[\begin{array}{c|c|c} \lambda 2 \ \partial_x\lambda 2 \ \partial_y\lambda 2 & 0 & 0 \\ 0 & \lambda 2 \ \partial_x\lambda 2 \ \partial_y\lambda 2 & 0 \\ 0 & 0 & \lambda 3_i \ \partial_x\lambda 3_i \ \partial_y\lambda 3_i \ \partial_{xx}\lambda 3_i \ \partial_{xy}\lambda 3_i \ \partial_{yy}\lambda 3_i \end{array} \right] \tag{3.4.5}$$

Following Figure 2.3.1, we need to use a numerical integration scheme *exact for polynomials of degree* 2 *on every sub-triangle* K_i *of the triangle* K. For instance, we can use the scheme given in Figure 2.2.1. Then, the relations (3.2.2) and (3.2.3) yield to

$$a_h(\vec{u_h},\vec{v_h}) = \sum_{K\in\mathcal{C}_h} \sum_{i=1}^{3} \left\{ [DG_i(\vec{u_h})]_{1\times 24} \; [M_i]_{24\times 24} \; {}^t[DG_i(\vec{v_h})]_{24\times 1} \right\} \quad (3.4.6)$$

where

$$\left.\begin{aligned} [M_i] = [DT_i]_{24\times 22} & \left\{ \text{aire } (K_i) \sum_{\ell_i=1}^{3} (\omega_{\ell_i} [LAMBD_i]_{22\times 12} \right. \\ & \left. [\mathbf{A}_{IJ}]_{12\times 12} \; {}^t[LAMBD_i])(b_{\ell_i}) \right\} {}^t[DT_i] \end{aligned}\right\} \quad (3.4.7)$$

and

$$f_h(\vec{v_h}) = \sum_{K\in\mathcal{C}_h} \sum_{i=1}^{3} \left\{ [B_i]_{1\times 24} \; {}^t[DG_i(\vec{v_h})]_{24\times 1} \right\} \quad (3.4.8)$$

with

$$\left.\begin{aligned} [B_i]_{1\times 24} = \text{aire } (K_i) & \left\{ \sum_{\ell_i=1}^{3} (\omega_{\ell_i} \; {}^t[\mathbf{F}]_{1\times 12} \; {}^t[LAMBD_i]_{12\times 22}) \right. \\ & \left. (b_{\ell_i}) \right\} {}^t[DT_i]_{22\times 24} \end{aligned}\right\} \quad (3.4.9)$$

<u>Remark 3.4.1</u> : A similar remark to Remark 3.2.1 applies to relations (3.4.6) and (3.4.8).

∎

<u>3.5. Energy functional and second member modules when the spaces X_{h1} and X_{h2} are constructed using reduced H.C.T. triangles</u> :

In a manner similar to that used in section 3.2, we obtain on every sub-triangle K_i of the triangle K :

$${}^t\mathbf{U}_h = [DG_i(\vec{u_h})]_{1\times 27} \; [DT_i]_{27\times 30} \; [LAMBD_i]_{30\times 12} \quad (3.5.1)$$

where, with the notations of relation (3.1.45), we have

$$[DG_i(\vec{u_h})]_{1\times 27} = [DLGL_{R_i}(u_{1h}) \; ; \; DLGL_{R_i}(u_{2h}) \; ; \; DLGL_{R_i}(u_{3h})] \quad (3.5.2)$$

$$[DT_i]_{27\times 30} = \begin{bmatrix} DA_{R_i} & 0 & 0 \\ 0 & DA_{R_i} & 0 \\ 0 & 0 & DA_{R_i} \end{bmatrix} \quad (3.5.3)$$

$$[DA_{R_i}]_{9\times 10} = [D_{R_i}]_{9\times 9} \; [A_{R_i}]_{9\times 10} \quad (3.5.4)$$

$$[LAMBD_i]_{30\times 12} =$$

$$\left[\begin{array}{c|c|c} \lambda 3_i \; \partial_x \lambda 3_i \; \partial_y \lambda 3_i & 0 & 0 \\ 0 & \lambda 3_i \; \partial_x \lambda 3_i \; \partial_y \lambda 3_i & 0 \\ 0 & 0 & \lambda 3_i \; \partial_x \lambda 3_i \; \partial_y \lambda 3_i \; \partial_{xx} \lambda 3_i \; \partial_{xy} \lambda 3_i \; \partial_{yy} \lambda 3_i \end{array}\right] \quad (3.5.5)$$

Following Figure 2.3.1, we need to use a numerical integration scheme *exact for polynomials of degree 4 on every sub-triangle* K_i *of the triangle* K. For example, we can use the scheme given in Figure 2.2.2. Then, the relations (3.2.2) and (3.2.3) yield to

$$a_h(\vec{u_h},\vec{v_h}) = \sum_{K\in\mathcal{T}_h} \sum_{i=1}^{3} \left\{ [DG_i(\vec{u_h})]_{1\times 27} \; [M_i]_{27\times 27} \; {}^t[DG_i(\vec{v_h})]_{27\times 1} \right\} \quad (3.5.6)$$

where

$$\left. \begin{array}{l} [M_i] = [DT_i]_{27\times 30} \left\{ \text{aire } (K_i) \sum_{\ell_i=1}^{6} (\omega_{\ell_i} [LAMBD_i]_{30\times 12} \right. \\ \left. [\mathbf{A}_{IJ}]_{12\times 12} \; {}^t[LAMBD_i]) \; (b_{\ell_i}) \right\} {}^t[DT_i] \end{array} \right\} \quad (3.5.7)$$

and

$$f_h(\vec{v_h}) = \sum_{K\in\mathcal{T}_h} \sum_{i=1}^{3} \left\{ [B_i]_{1\times 27} \; {}^t[DG_i(\vec{v_h})]_{27\times 1} \right\} \quad (3.5.8)$$

$$[B_i]_{1\times 27} = \text{aire } (K_i) \left\{ \sum_{\ell_i=1}^{6} (\omega_{\ell_i} \; {}^t[\mathbf{F}]_{1\times 12} \; {}^t[LAMBD_i]_{12\times 30})(b_{\ell_i}) \right\} {}^t[DT_i]_{30\times 27}$$

(3.5.9)

Remark 3.5.1 : A similar remark to Remark 3.2.1 applies to relations (3.5.6) and (3.5.8). ∎

3.6. Energy functional and second member modules when the spaces X_{h1} and X_{h2} are constructed using triangles of type (1) and reduced H.C.T. triangles, respectively :

Using the strategy of section 3.2, we obtain on every sub-triangle K_i of the triangle K :

$$ {}^t\mathbf{U}_h = [DG_i(\vec{u_h})]_{1\times 15} \; [DT_i]_{15\times 16} \; [LAMBD_i]_{16\times 12} \tag{3.6.1}$$

where, with the notations of relations (3.1.6) and (3.1.45), we have

$$[DG_i(\vec{u_h})]_{1\times 15} = [DL_1(u_{1h}) \; ; \; DL_1(u_{2h}) \; ; \; DLGL_{R_i}(u_{3h})] \; , \tag{3.6.2}$$

$$[DT_i]_{15\times 16} = \begin{bmatrix} I_3 & 0 & 0 \\ 0 & I_3 & 0 \\ 0 & 0 & DA_{R_i} \end{bmatrix} \tag{3.6.3}$$

$$[DA_{R_i}]_{9\times 10} = [D_{R_i}]_{9\times 9} \; [A_{R_i}]_{9\times 10} \; , \tag{3.6.4}$$

$$[LAMBD_i]_{16\times 12} =$$

$$\left[\begin{array}{c|c|c} \lambda 1 \;\; \partial_x\lambda 1 \;\; \partial_y\lambda 1 & 0 & 0 \\ 0 & \lambda 1 \;\; \partial_x\lambda 1 \;\; \partial_y\lambda 1 & 0 \\ 0 & 0 & \lambda 3_i \;\; \partial_x\lambda 3_i \;\; \partial_y\lambda 3_i \;\; \partial_{xx}\lambda 3_i \;\; \partial_{xy}\lambda 3_i \;\; \partial_{yy}\lambda 3_i \end{array}\right] \tag{3.6.5}$$

Following Figure 2.3.1, we need to use a numerical integration scheme *exact for polynomials of degree* 2 *on every sub-triangle* K_i *of the triangle* K. For instance, we can use the scheme given in Figure 2.2.1. Then, the relations (3.2.2) and (3.2.3) yield to

$$a_h(\vec{u_h},\vec{v_h}) = \sum_{K\in\mathcal{C}_h} \sum_{i=1}^{3} \left\{ [DG_i(\vec{u_h})]_{1\times 15} \; [M_i]_{15\times 15} \; {}^t[DG_i(\vec{v_h})]_{15\times 1} \right\} \tag{3.6.6}$$

where

$$[M_i]_{15\times15} = [DT_i]_{15\times16} \left\{ \text{aire}\ (K_i) \sum_{\ell_i=1}^{3} (\omega_{\ell_i} [LAMBD_i]_{16\times12} \right. \\ \left. [\mathbf{A}_{IJ}]_{12\times12}\ {}^t[LAMBD_i]\ (b_{\ell_i}) \right\}\ {}^t[DT_i] \tag{3.6.7}$$

and

$$f_h(\overrightarrow{v_h}) = \sum_{K\in\mathscr{C}_h} \sum_{i=1}^{3} \left\{ [B_i]_{1\times15}\ {}^t[DG_i(\overrightarrow{v_h})]_{15\times1} \right\} \tag{3.6.8}$$

with

$$[B_i]_{1\times15} = \text{aire}\ (K_i) \left\{ \sum_{\ell_i=1}^{3} (\omega_{\ell_i}\ {}^t[\mathbf{F}]_{1\times12}\ {}^t[LAMBD_i]_{12\times16}) \right. \\ \left. (b_{\ell_i}) \right\}\ {}^t[DT_i]_{16\times15}\ . \tag{3.6.9}$$

<u>Remark 3.6.1</u> : A similar remark to Remark 3.2.1 applies to relations (3.6.6) and (3.6.8).

■

PART II : APPLICATION TO ARCH DAM SIMULATIONS

Introduction :

In order to illustrate the developments of Part I, we describe in Part II an application to the problem of an *arch dam* subjected to

- its *weight*
- *hydrostatic pressure*, and to
- *changes of temperatures*.

For definiteness, we consider the example of the arch version of the project of the GRAND'MAISON dam studied by "COYNE et BELLIER" [1977] for "Electricité de France". For technical and financial considerations, the earth dam version has been finally preferred and constructed. By analogy, it would be possible to consider other types of arch dams.

In all of this application, our primary interest is in the numerical aspects of the problem and not in design features or physical behavior of the dam. However, the numerical experiments can be qualitatively compared with the experimental results given in NAYLOR, STAGG, and ZIENKIEWICZ [1975] or in RYDZEWSKI [1965] for similar arch dams. In order to slightly improve the results, we could consider some additional effects, particularly the movement of foundations as well as shrinkage, creep, earthquake, dynamical behaviour,... . Another interesting study would be the optimization of the arch dam thickness so that the concrete volume is decreased without danger. For an introduction to some of these different aspects of arch dam analysis, we again refer for instance to NAYLOR, STAGG and ZIENKIEWICZ [1975].

In Chapter 4, we recall the *definition of the dam* as given by "COYNE et BELLIER". This definition appears also in Figures 4.1.1 to 4.1.5. So, with the help of curvilinear coordinates, we define the middle surface and the thickness of the dam.

Next, in Chapter 5, we obtain the expression of the potential energy of external loads using resultants on the middle surface of the dam. Thus, we derive a *variational formulation of the continuous problem.*

In Chapter 6, *we describe the implementation* of the corresponding discrete problems. In particular, we detail the method of triangulation and the way to take into account the boundary conditions and the symmetry conditions. Thus, we define discrete problems on the half-reference domain. Corresponding linear systems are solved by a *CHOLESKI method* using the *sky-line bandwidth factorization*. From the knowledge of an approximation of the displacement field we obtain approximation of the *physical components of the stresses.*

Finally, some significant numerical results are given in Chapter 7.

CHAPTER 4

GEOMETRICAL DEFINITION OF THE DAM

Orientation :

In this chapter we record the *geometrical definition of the dam* of the pre-project of GRAND'MAISON arch dam proposed by COYNE-et-BELLIER [1977] at the request of ELECTRICITE DE FRANCE. Since the thickness of the dam is small with respect to the two other dimensions, we shall represent the dam as a thin shell in the sense of KOITER [1966, 1970]. In this goal, we give the definition of the middle surface of the dam as also the definitions of the corresponding geometrical parameters. The meaning of these parameters is specified in BERNADOU [1978, 1980] or in BERNADOU-CIARLET [1976].

4.1. A pre-project of a dam

Figures 4.1.1 to 4.1.5 are extracted from the preliminary file of GRAND'MAISON arch dam studied by COYNE-et-BELLIER [1977]. Figures 4.1.1 and 4.1.2 give the *site plane of the dam* with different level lines as also excavation results. In Figures 4.1.3 to 4.1.5, we have the *geometrical definition of the dam*, horizontal sections of the upper and lower arcs as also some vertical sections.

In these figures one can note that the definition of this arch dam is very close to the definition of a thin shell as given in (1.2.2). In order to use the results of BERNADOU [1978, 1980] and BERNADOU and BOISSERIE [1978a, 1978b] we introduce two slight modifications to the geometrical definition of the dam as proposed by COYNE-et-BELLIER :

(1) *the middle surface of the dam will be assumed symmetric* with respect to the vertical plane including the dam axis (see section 4.2). This is not true in Figure 4.1.3 : at the heights 1697 and 1630, lateral

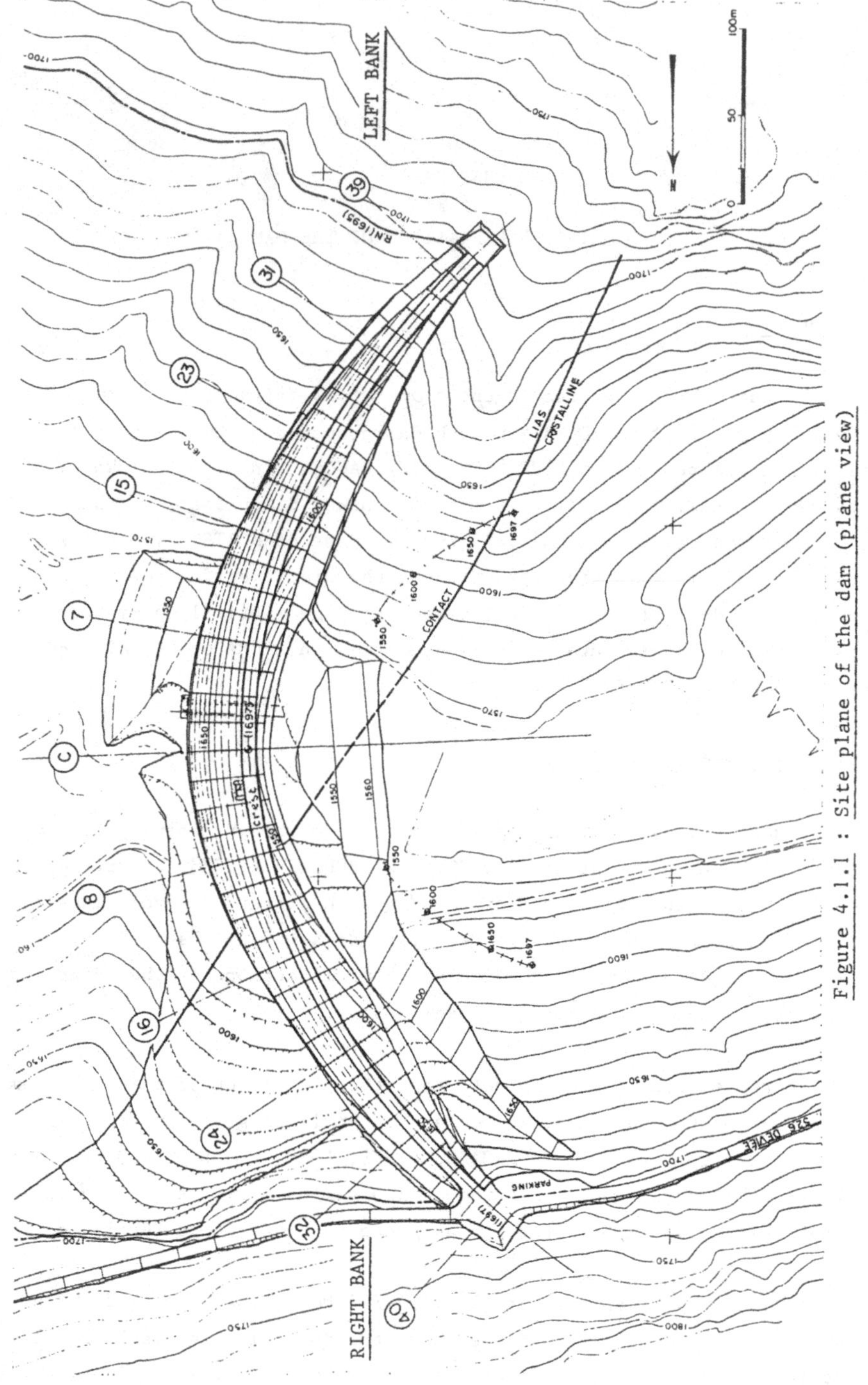

Figure 4.1.1 : Site plane of the dam (plane view)

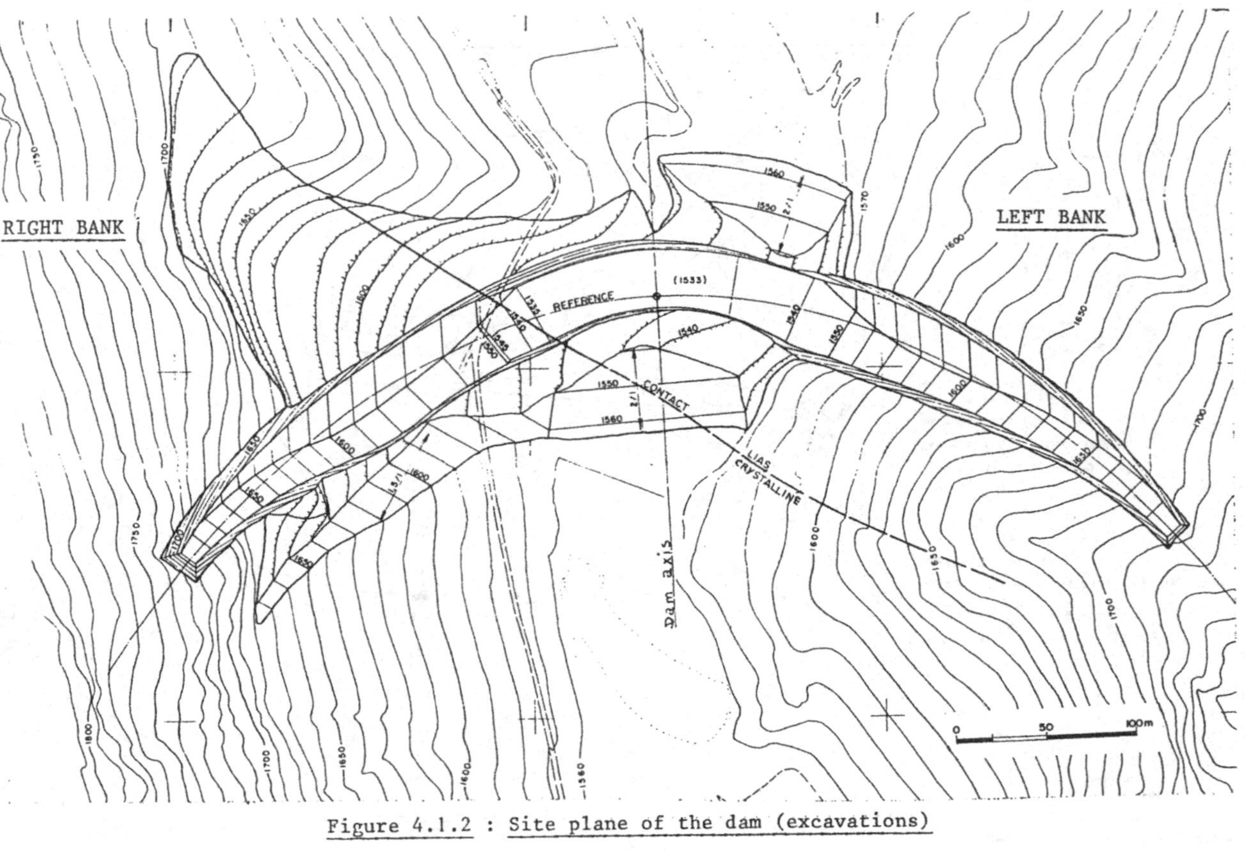

Figure 4.1.2 : Site plane of the dam (excavations)

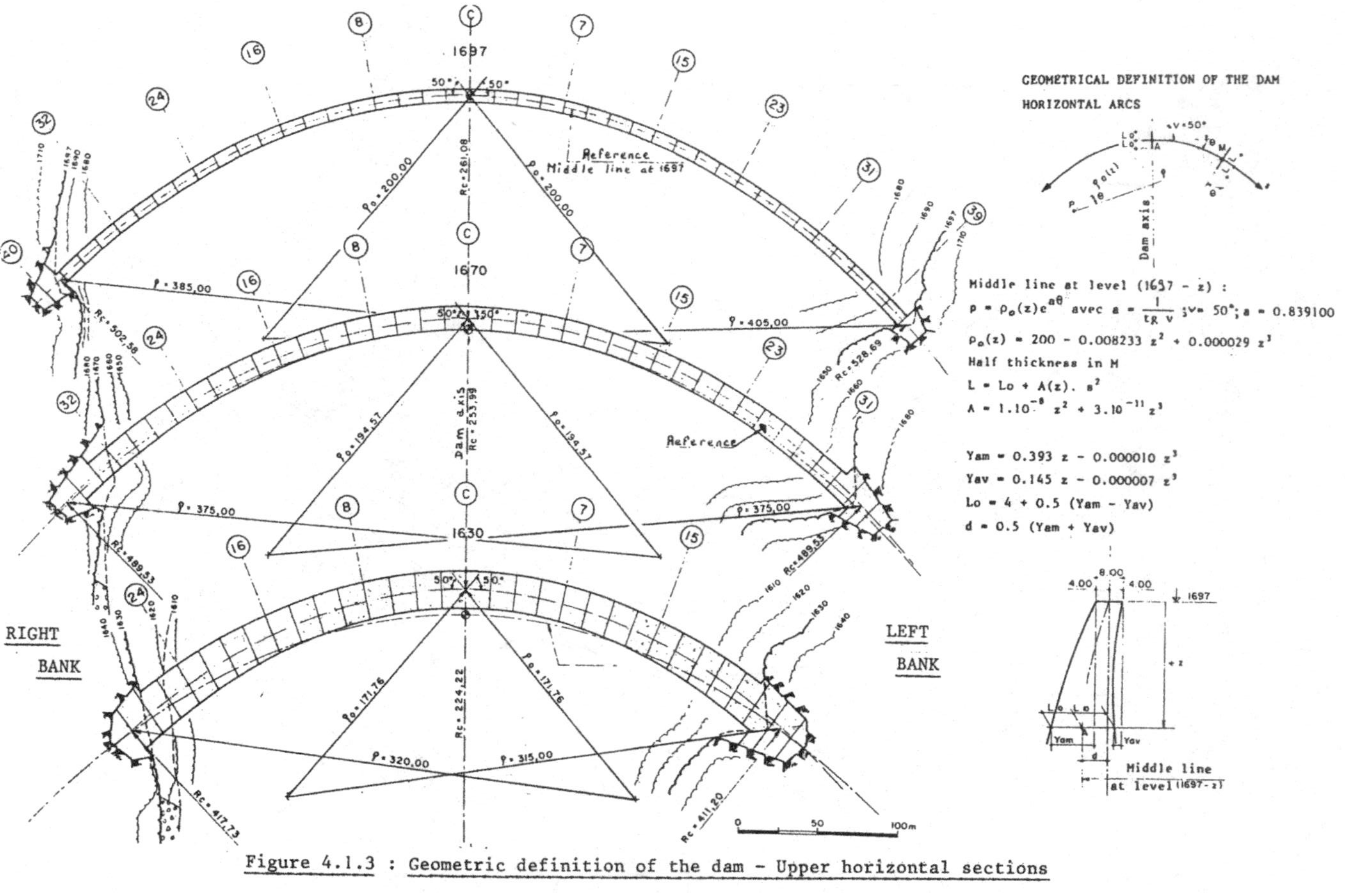

Figure 4.1.3 : Geometric definition of the dam - Upper horizontal sections

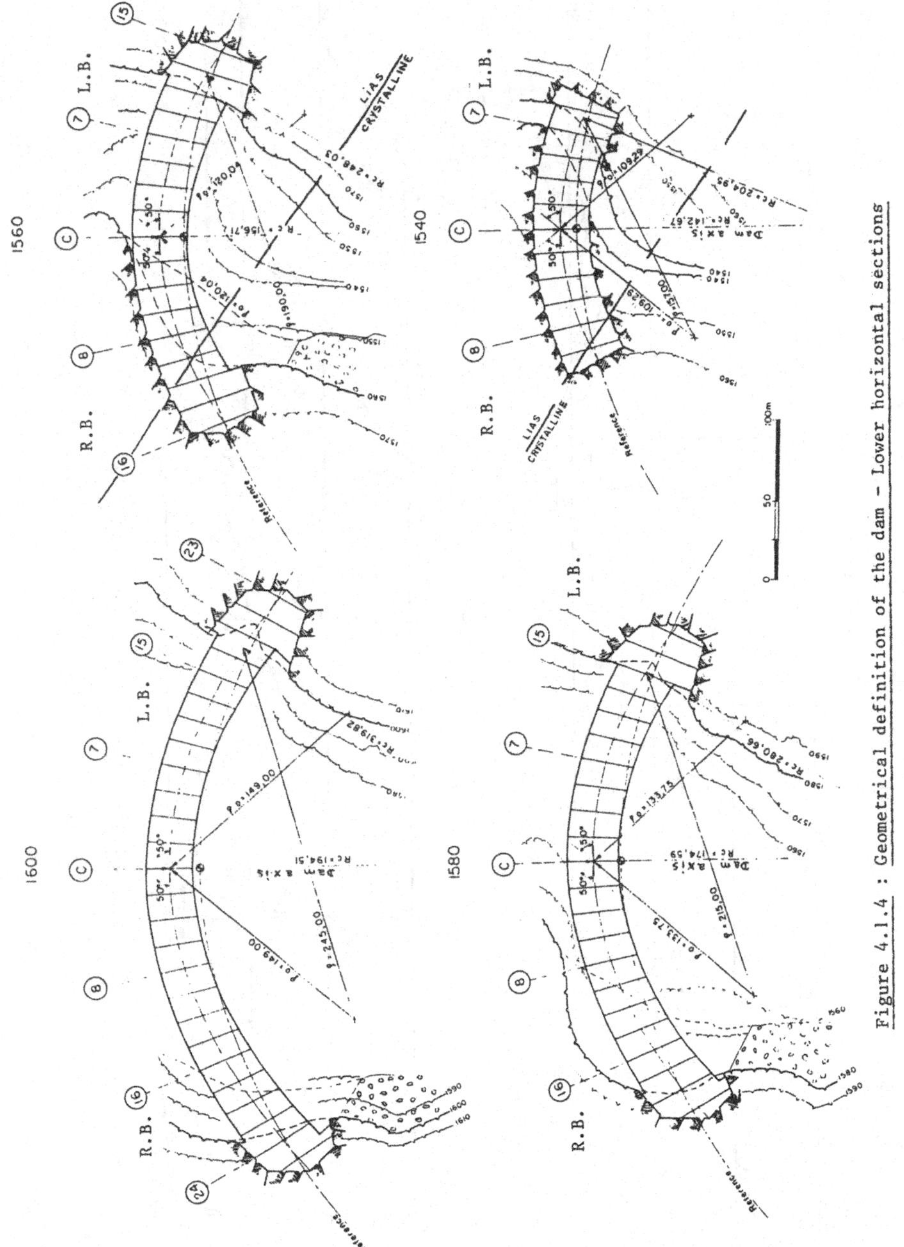

Figure 4.1.4 : Geometrical definition of the dam - Lower horizontal sections

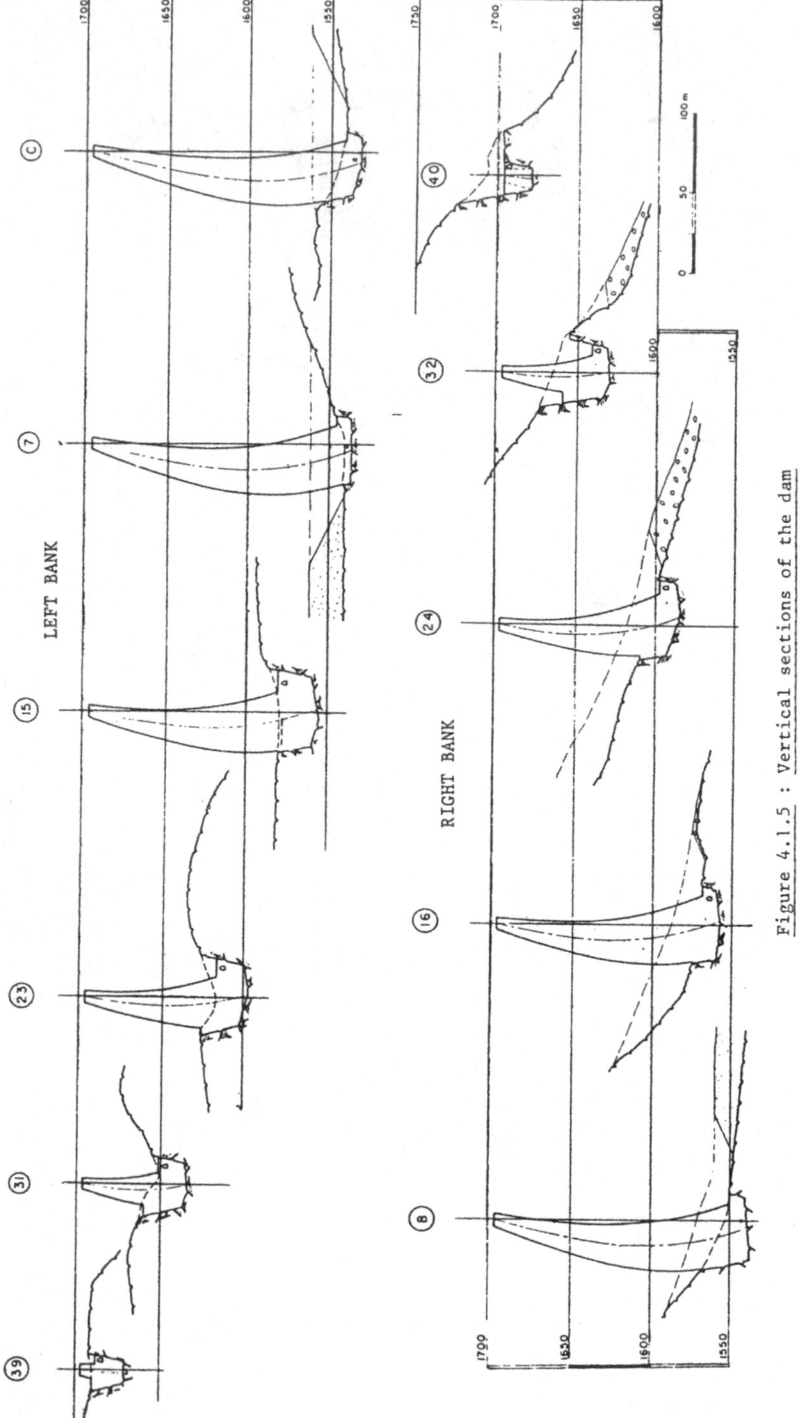

Figure 4.1.5 : Vertical sections of the dam

sides are not symmetrical.

(2) *the thickness of the shell will be measured along the normal to the middle surface* - see section 4.4. - in agreement with definition (1.2.2). This is not true in Figure 4.1.3 where the thickness is measured along an horizontal line.

4.2. Definition of the middle surface

The middle surface S of the arch dam is referred to a fixed orthonormal reference system $(0, \vec{e_1}, \vec{e_2}, \vec{e_3})$ of the euclidean space $\mathcal{E}^3$ as indicated in Figure 4.2.1. The origin of this reference system is located at the intersection of the crest line with the symmetry plane of the arch dam. Vector $\vec{e_1}$ is horizontal, in the symmetry plane of the arch dam and oriented towards the upstream part of the dam. Vector $\vec{e_3}$ is vertical and oriented from top to bottom. In this reference system, the coordinates (x^1, x^2, x^3) of any point M are such that

$$\vec{OM} = x^i \vec{e_i} \quad , \tag{4.2.1}$$

(using EINSTEIN's convention).

The middle surface S of the arch dam is defined as the image of a plane domain Ω through a map $\vec{\phi}$, i.e.,

$$\vec{\phi} : (\xi^1,\xi^2) \in \Omega \mapsto \vec{OM} = \vec{\phi}(\xi^1,\xi^2) = x^i(\xi^1,\xi^2)\vec{e_i} \quad . \tag{4.2.2}$$

The curvilinear coordinates (ξ^1,ξ^2) are given by

$$(\xi^1 = \frac{\theta}{\theta_\circ} \quad , \quad \xi^2 = \frac{Z}{Z_\circ}) \quad , \tag{4.2.3}$$

where the parameters θ and Z are defined in Figure 4.1.3 and where $\theta_\circ = \max|\theta|$, $Z_\circ = \max Z$: the maximum is taken along the arch dam, for instance $Z_\circ$ is the height of the arch dam.

Then, by observing Figure 4.1.3, we see that the generic point P of the middle surface S is defined by its coordinates :

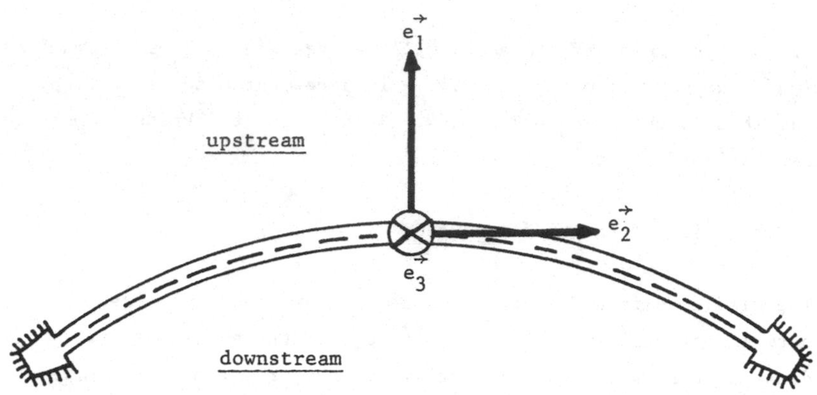

Crest line of the arch dam

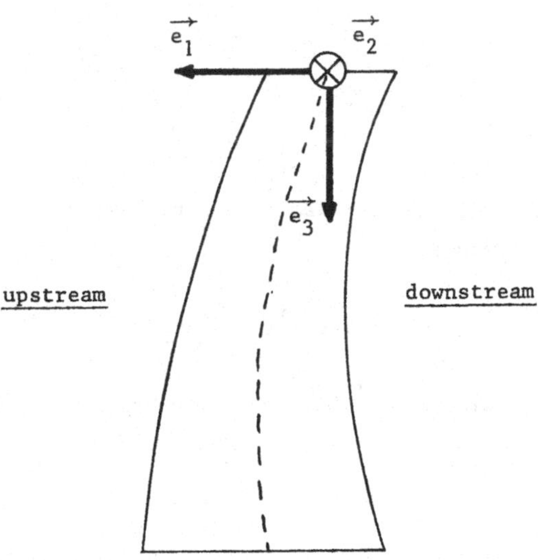

Figure 4.2.1 : The fixed orthonormal reference system of the space $\mathcal{E}^3$ (In the figure, the thickness of the dam is defined horizontally as indicated in COYNE and BELLIER [1977]).

$$\left.\begin{aligned} x^1(\xi^1,\xi^2) &= \rho_\circ(\xi^2)\,[e^{\alpha\theta_\circ|\xi^1|}\cos(\theta_\circ|\xi^1|+40^\circ) - \cos 40^\circ] \\ &\quad + 0.269\,Z_\circ\xi^2 - 0.0000085\,Z_\circ^3(\xi^2)^3 \\ x^2(\xi^1,\xi^2) &= \frac{|\xi^1|}{\xi^1}\,\rho_\circ(\xi^2)\,[e^{\alpha\theta_\circ|\xi^1|}\sin(\theta_\circ|\xi^1|+40^\circ) - \sin 40^\circ] \\ x^3(\xi^1,\xi^2) &= Z_\circ\xi^2 \end{aligned}\right\} \quad (4.2.4)$$

where constants α, $\theta_\circ$, $Z_\circ$ and function $\rho_\circ(\xi^2)$ are defined by

$$\left.\begin{aligned} \alpha &= \text{tg } 40^\circ \\ \theta_\circ &= 48^\circ\ 178 \\ Z_\circ &= 157 \end{aligned}\right\} \quad (4.2.5)$$

$$\rho_\circ(\xi^2) = 200 - 0.008233(Z_\circ)^2(\xi^2)^2 + 0.000029(Z_\circ)^3(\xi^2)^3 \quad (4.2.6)$$

In order to complete the definition of the reference domain Ω, it remains to specify the bounds of the intervals of variation of the parameter θ as function of Z. For every horizontal section given in Figures 4.1.3 and 4.1.4, corresponding bounds are indicated in Table 4.2.1. Then, by using a linear interpolation of the boundary from an horizontal section to the following, we obtain the reference domain $\tilde{\Omega}$ indicated in Figure 4.2.2.

In the following, it will be convenient to approximate the reference domain $\tilde{\Omega}$ by a symmetric trapezoïdal domain Ω. This simplification does not involve any loss of generality and it allows to work only on the half-domain $(\xi^1,\xi^2) \in \Omega$; $\xi^1 > 0$.

		Right limit ($\xi^1 > 0$)		Left limit ($\xi^1 < 0$)	
Z = 0	$\rho_\circ$ = 200	ρ = 405	$\theta = 48^\circ\ 178$	ρ = 385	$\theta = -\ 44^\circ\ 720$
Z = 27	$\rho_\circ$ = 194.57	ρ = 375	$\theta = 44^\circ\ 802$	ρ = 375	$\theta = -\ 44^\circ\ 802$
Z = 67	$\rho_\circ$ = 171.76	ρ = 315	$\theta = 41^\circ\ 412$	ρ = 320	$\theta = -\ 42^\circ\ 487$
Z = 97	$\rho_\circ$ = 149.00	ρ = 245	$\theta = 33^\circ\ 958$	ρ = 245	$\theta = -\ 33^\circ\ 958$
Z = 117	$\rho_\circ$ = 133.75	ρ = 215	$\theta = 32^\circ\ 411$	ρ = 215	$\theta = -\ 32^\circ\ 411$
Z = 137	$\rho_\circ$ = 120.04	ρ = 190	$\theta = 31^\circ\ 355$	ρ = 190	$\theta = -\ 31^\circ\ 355$
Z = 157	$\rho_\circ$ = 109.29	ρ = 157	$\theta = 24^\circ\ 735$	ρ = 157	$\theta = -\ 24^\circ\ 735$

Table 4.2.1 : Values of the parameters at the limits of horizontal sections ($\theta = \frac{\text{Log } \rho - \text{Log } \rho_\circ}{\alpha}$)

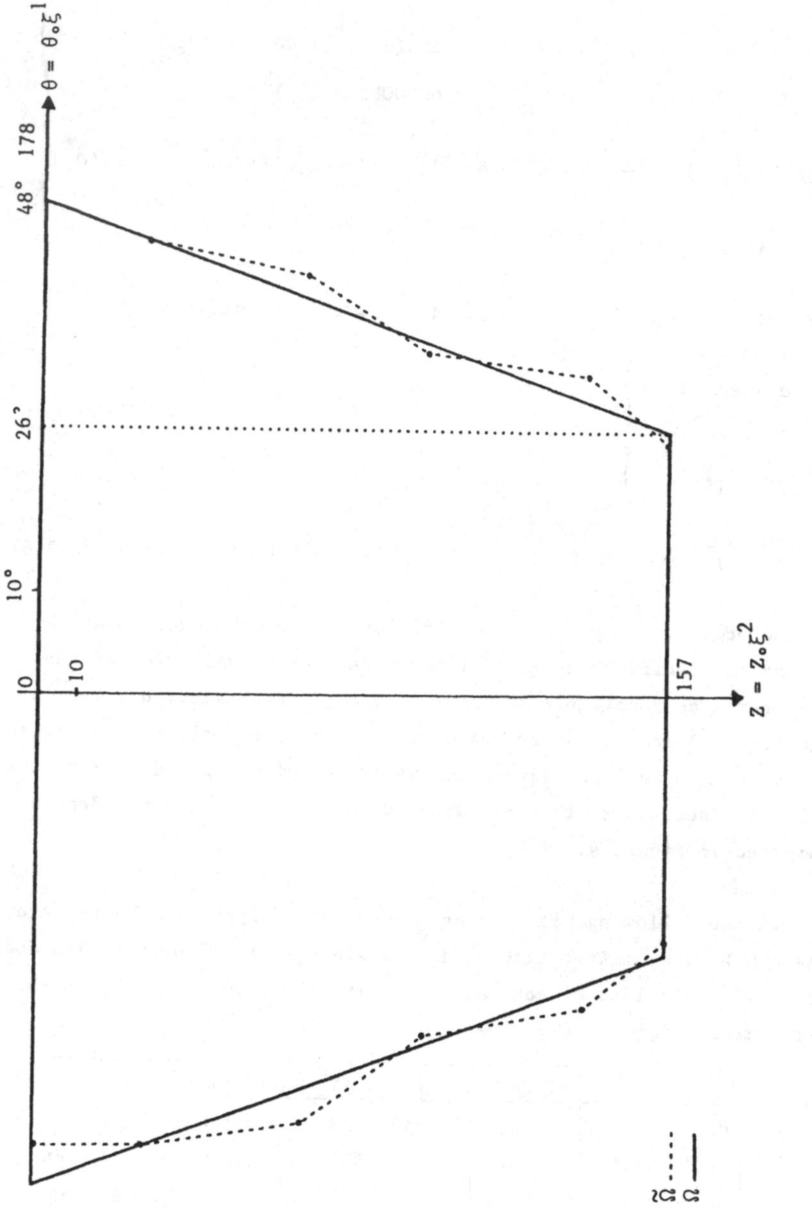

Figure 4.2.2 : Reference domains $\tilde{\Omega}$ and Ω

4.3. Calculation of the geometrical parameters of the middle surface :

We have observed in section 1.5 that the geometrical parameters which appear in calculation of the strain energy depend on first, second and third partial derivatives of the mapping $\vec{\phi}$. Relations (4.2.4) involve :

First partial derivatives

$$\vec{\phi}_{,1} = \vec{a_1} = \frac{\theta_\circ e^{\alpha\theta_\circ|\xi^1|}\rho_\circ(\xi^2)}{\cos 40^\circ}\begin{cases} -\frac{|\xi^1|}{\xi^1}\sin(\theta_\circ|\xi^1|) \\ \cos(\theta_\circ|\xi^1|) \\ 0 \end{cases} \qquad (4.3.1)$$

$$\vec{\phi}_{,2} = \vec{a_2} = \begin{cases} \rho_\circ'(\xi^2)\,[e^{\alpha\theta_\circ|\xi^1|}\cos(\theta_\circ|\xi^1| + 40^\circ) - \cos 40^\circ] + \\ \qquad + 0.269\, Z_\circ - 0.0000255(Z_\circ)^3(\xi^2)^2 \\ \frac{|\xi^1|}{\xi^1}\rho_\circ'(\xi^2)\,[e^{\alpha\theta_\circ|\xi^1|}\sin(\theta_\circ|\xi^1| + 40^\circ) - \sin 40^\circ] \\ Z_\circ \end{cases} \qquad (4.3.2)$$

where

$$\rho_\circ'(\xi^2) = -\,0.016466(Z_\circ)^2\,\xi^2 + 0.000087(Z_\circ)^3(\xi^2)^2 \qquad (4.3.3)$$

Second partial derivatives

$$\vec{\phi}_{,11} = -\frac{(\theta_\circ)^2 e^{\alpha\theta_\circ|\xi^1|}\rho_\circ(\xi^2)}{(\cos 40^\circ)^2}\begin{cases} \cos(\theta_\circ|\xi^1| - 40^\circ) \\ \frac{|\xi^1|}{\xi^1}\sin(\theta_\circ|\xi^1| - 40^\circ) \\ 0 \end{cases} \qquad (4.3.4)$$

$$\vec{\phi}_{,12} = \frac{\theta_\circ e^{\alpha\theta_\circ|\xi^1|}\rho_\circ'(\xi^2)}{\cos 40^\circ}\begin{cases} -\frac{|\xi^1|}{\xi^1}\sin(\theta_\circ|\xi^1|) \\ \cos(\theta_\circ|\xi^1|) \\ 0 \end{cases} \qquad (4.3.5)$$

$$\vec{\phi}_{,22} = \begin{cases} \rho_\circ''(\xi^2)\,[e^{\alpha\theta_\circ|\xi^1|}\cos(\theta_\circ|\xi^1| + 40^\circ) - \cos 40^\circ] - \\ \qquad - 0.000051(Z_\circ)^3\xi^2 \\ \frac{|\xi^1|}{\xi^1}\cdot\rho_\circ''(\xi^2)\,[e^{\alpha\theta_\circ|\xi^1|}\sin(\theta_\circ|\xi^1| + 40^\circ) - \sin 40^\circ] \\ 0 \end{cases} \qquad (4.3.6)$$

where

$$\rho_\circ''(\xi^2) = -\ 0.016466(Z_\circ)^2 + 0.000174(Z_\circ)^3\ \xi^2\ . \qquad (4.3.7)$$

<u>Third partial derivatives</u>

$$\vec{\phi}_{,111} = \frac{(\theta_\circ)^3 e^{\alpha\theta_\circ|\xi^1|}\rho_\circ(\xi^2)}{(\cos 40^\circ)^3} \begin{cases} \frac{|\xi^1|}{\xi^1}\sin(\theta_\circ|\xi^1| - 80^\circ) \\ -\cos(\theta_\circ|\xi^1| - 80^\circ) \\ 0 \end{cases} \qquad (4.3.8)$$

$$\vec{\phi}_{,112} = \frac{-(\theta_\circ)^2 e^{\alpha\theta_\circ|\xi^1|}\rho_\circ'(\xi^2)}{(\cos 40^\circ)^2} \begin{cases} \cos(\theta_\circ|\xi^1| - 40^\circ) \\ \frac{|\xi^1|}{\xi^1}\sin(\theta_\circ|\xi^1| - 40^\circ) \\ 0 \end{cases} \qquad (4.3.9)$$

$$\vec{\phi}_{,122} = \frac{\theta_\circ e^{\alpha\theta_\circ|\xi^1|}\rho_\circ''(\xi^2)}{\cos 40^\circ} \begin{cases} -\frac{|\xi^1|}{\xi^1}\sin(\theta_\circ|\xi^1|) \\ \cos(\theta_\circ|\xi^1|) \\ 0 \end{cases} \qquad (4.3.10)$$

$$\vec{\phi}_{,222} = \begin{cases} \rho_\circ'''(\xi^2)\ [e^{\alpha\theta_\circ|\xi^1|}\cos(\theta_\circ|\xi^1| + 40^\circ) - \cos 40^\circ] - \\ \qquad - 0{,}000051(Z_\circ)^3 \\ \frac{|\xi^1|}{\xi^1}\rho_\circ'''(\xi^2)\ [e^{\alpha\theta_\circ|\xi^1|}\sin(\theta_\circ|\xi^1| + 40^\circ) - \sin 40^\circ] \\ 0 \end{cases} \qquad (4.3.11)$$

where

$$\rho_\circ'''(\xi^2) = 0.000174(Z_\circ)^3 \qquad (4.3.12)$$

In the above expressions, we have only mentionned the geometrical parameters which appear effectively in calculations (see relations (1.5.12) to (1.5.18)). Nevertheless, for checking purposes, it is interesting to dispose of analytical values of two fundamental forms, of the expression giving the *curvature radius* R_C which appears in Figures 4.1.3 and 4.1.4 as well as of CHRISTOFFEL's symbols. Thus we obtain successively :

First fundamental form (see (1.1.4) and (1.5.12)) :

$$
\left.
\begin{aligned}
a_{11} &= \frac{(\theta_\circ)^2 e^{2\alpha\theta_\circ|\xi^1|}(\rho_\circ(\xi^2))^2}{(\cos 40^\circ)^2} \\
a_{12} = a_{21} &= \frac{|\xi^1|\theta_\circ e^{\alpha\theta_\circ|\xi^1|}\rho_\circ(\xi^2)}{\xi^1 \cos 40^\circ} [\rho_\circ'(\xi^2) \left\{ e^{\alpha\theta_\circ|\xi^1|} \sin 40^\circ + \right. \\
&\left. + \sin(\theta_\circ|\xi^1| - 40^\circ) \right\} - \left\{ 0.269 Z_\circ - 0.0000255 (Z_\circ)^3 (\xi^2)^2 \right\} \\
&\times \sin(\theta_\circ|\xi^1|)] \\
a_{22} &= (\rho_\circ'(\xi^2))^2 \, (e^{2\alpha\theta_\circ|\xi^1|} - 2e^{\alpha\theta_\circ|\xi^1|}\cos(\theta_\circ|\xi^1|) + 1) \\
&+ 2\rho_\circ'(\xi^2) \, [e^{\alpha\theta_\circ|\xi^1|} \cos(\theta_\circ|\xi^1| + 40^\circ) - \\
&- \cos 40^\circ] \, [0.269 Z_\circ - 0.0000255 (Z_\circ)^3 (\xi^2)^2] \\
&+ [0.269 Z_\circ - 0.0000255 (Z_\circ)^3 (\xi^2)^2]^2 + (Z_\circ)^2
\end{aligned}
\right\} \quad (4.3.13)
$$

$$
\left.
\begin{aligned}
a = \frac{(\theta_\circ)^2 e^{2\alpha\theta_\circ|\xi^1|}(\rho_\circ(\xi^2))^2}{(\cos 40^\circ)^2} & \left\{ (Z_\circ)^2 + \right. \\
&+ \{\rho_\circ'(\xi^2) \, [e^{\alpha\theta_\circ|\xi^1|} \cos 40^\circ - \cos(\theta_\circ|\xi^1| - 40^\circ)] + \\
&\left. + [0.269 Z_\circ - 0.0000255 (Z_\circ)^3 (\xi^2)^2] \cos(\theta_\circ|\xi^1|)\}^2 \right\}
\end{aligned}
\right\} \quad (4.3.14)
$$

Second fundamental form (see (1.1.5) and (1.5.15)) :

$$
\left.
\begin{aligned}
b_{11} &= - \frac{(\theta_\circ)^3}{\sqrt{a}} \frac{e^{2\alpha\theta_\circ|\xi^1|}(\rho_\circ(\xi^2))^2}{(\cos 40^\circ)^2} Z_\circ \\
b_{12} &= b_{21} = 0 \\
b_{22} &= \frac{\theta_\circ Z_\circ}{\sqrt{a}} \frac{e^{\alpha\theta_\circ|\xi^1|}\rho_\circ(\xi^2)}{\cos 40^\circ} [\rho_\circ''(\xi^2) \left\{ e^{\alpha\theta_\circ|\xi^1|} \cos 40^\circ - \right. \\
&\left. - \cos(\theta_\circ|\xi^1| - 40^\circ) \right\} - 0.000051 (Z_\circ)^3 \, \xi^2 \cos(\theta_\circ|\xi^1|)]
\end{aligned}
\right\} \quad (4.3.15)
$$

Curvature radius R_C of the meridian line :

We now consider the question of evaluating the radius of curvature of the meridian line described by point $M(\xi^1,\xi^2)$ when ξ^1 varies and $\xi^2 \doteq$ constant. Denoting by $\vec{t}$ the unit tangent vector to this meridian line, relation (4.3.1) yields

$$\vec{t} = \begin{cases} -\dfrac{\xi^1}{|\xi^1|}\sin(\theta_\circ|\xi^1|) \\ \cos(\theta_\circ|\xi^1|) \\ 0 \end{cases}$$

from which we deduce the unit normal vector $\vec{n}$ to this same meridian line, i.e.,

$$\vec{n} = \vec{t} \times \vec{e_3} = \begin{cases} \cos(\theta_\circ|\xi^1|) \\ \dfrac{\xi^1}{|\xi^1|}\sin(\theta_\circ|\xi^1|) \\ 0 \end{cases}$$

But, from the FRENET-SERRET formulae, we get

$$\frac{\vec{dt}}{ds} = \frac{\vec{n}}{R_C}$$

where ds denotes the line element of the meridian line. For ξ^2 = constant, we find

$$(ds)^2 = (dx^1)^2 + (dx^2)^2 \quad ,$$

hence, with (4.2.4) and (4.3.1) :

$$ds = \frac{\theta_\circ e^{\alpha\theta_\circ|\xi^1|}\rho_\circ(\xi^2)}{\cos 40^\circ}\, d\xi^1 \quad . \tag{4.3.16}$$

Thus, we obtain

$$R_C = -\frac{e^{\alpha\theta_\circ|\xi^1|}\rho_\circ(\xi^2)}{\cos 40^\circ} \quad . \tag{4.3.17}$$

<u>CHRISTOFFEL's symbols</u> $\Gamma_{\alpha\beta\gamma} = \Gamma_{\alpha\gamma\beta} = \vec{a}_\alpha \cdot \vec{a}_{\beta,\gamma}$:

$$\Gamma_{111} = \vec{\phi}_{,1} \cdot \vec{\phi}_{,11} = \frac{(\theta_\circ)^3 e^{2\alpha\theta_\circ|\xi^1|}(\rho_\circ(\xi^2))^2}{(\cos 40^\circ)^3}\,\frac{|\xi^1|}{\xi^1}\sin 40^\circ \tag{4.3.18}$$

$$\Gamma_{112} = \vec{\phi}_{,1}\cdot\vec{\phi}_{,12} = \frac{(\theta_\circ)^2 e^{2\alpha\theta_\circ|\xi^1|}\rho_\circ(\xi^2)\rho_\circ'(\xi^2)}{(\cos 40^\circ)^2} \tag{4.3.19}$$

$$\left.\begin{aligned}\Gamma_{122} &= \vec{\phi}_{,1}\cdot\vec{\phi}_{,22} = \frac{|\xi^1|\theta_\circ e^{\alpha\theta_\circ|\xi^1|}\rho_\circ(\xi^2)}{\xi^1 \cos 40^\circ} \\ &\times [\rho_\circ''(\xi^2) \left\{ e^{\alpha\theta_\circ|\xi 1|} \sin 40^\circ + \sin(\theta_\circ|\xi^1| - 40^\circ)\right\} \\ &+ 0.000051(Z_\circ)^3\ \xi^2 \sin(\theta_\circ|\xi^1|)]\end{aligned}\right\} \tag{4.3.20}$$

$$\left.\begin{aligned}\Gamma_{211} &= \vec{\phi}_{,2}\cdot\vec{\phi}_{,11} = -\frac{(\theta_\circ)^2 e^{\alpha\theta_\circ|\xi^1|}\rho_\circ(\xi^2)}{(\cos 40^\circ)^2} \\ &\times [\rho_\circ'(\xi^2) \left\{e^{\alpha\theta_\circ|\xi^1|} \cos 80^\circ - \cos(\theta_\circ|\xi^1| - 80^\circ)\right\} \\ &+ \left\{0.269Z_\circ - 0.0000255(Z_\circ)^3\ (\xi^2)^2\right\} \cos(\theta_\circ|\xi^1| - 40^\circ)]\end{aligned}\right\} \tag{4.3.21}$$

$$\left.\begin{aligned}\Gamma_{212} &= \vec{\phi}_{,2}\cdot\vec{\phi}_{,12} = \frac{|\xi^1|}{\xi^1}\,\frac{\theta_\circ e^{\alpha\theta_\circ|\xi^1|}\rho_\circ'(\xi^2)}{\cos 40^\circ} \\ &\times [\rho_\circ'(\xi^2) \left\{e^{\alpha\theta_\circ|\xi^1|} \sin 40^\circ + \sin(\theta_\circ|\xi^1| - 40^\circ)\right\} \\ &- \left\{0.269Z_\circ - 0.0000255(Z_\circ)^3\ (\xi^2)^2\right\} \sin(\theta_\circ|\xi^1|)]\end{aligned}\right\} \tag{4.3.22}$$

$$\left.\begin{aligned}\Gamma_{222} &= \vec{\phi}_{,2}\cdot\vec{\phi}_{,22} = \rho_\circ'(\xi^2)\ \rho_\circ''(\xi^2)\ [e^{2\alpha\theta_\circ|\xi^1|} - \\ &- 2e^{\alpha\theta_\circ|\xi^1|} \cos(\theta_\circ|\xi^1|) + 1] \\ &+ \left\{e^{\alpha\theta_\circ|\xi^1|} \cos(\theta_\circ|\xi^1| + 40^\circ) - \cos 40^\circ\right\} \\ &\times \left\{(0.269Z_\circ - 0.0000255(Z_\circ)^3\ (\xi^2)^2)\ \rho_\circ''(\xi^2)\right. \\ &\left.- 0.000051(Z_\circ)^3\ \xi^2\ \rho_\circ'(\xi^2)\right\} \\ &- [0.269Z_\circ - 0.0000255(Z_\circ)^3(\xi^2)^2]\ 0.000051(Z_\circ)^3\ \xi^2\end{aligned}\right\} \tag{4.3.23}$$

from which we deduce the expressions of coefficients $\Gamma^\alpha_{\beta\gamma}$ by using relations :

$$\Gamma^{\alpha}_{\beta\gamma} = a^{\alpha\lambda} \, \Gamma_{\lambda\beta\gamma} \, . \qquad (4.3.24)$$

Remark 4.3.1 : Relations (4.3.1) to (4.3.24) show that :

(i) *the following functions are continuous in* $\bar{\Omega}$:

$$\vec{\phi}_{,\alpha} \; ; \; \vec{\phi}_{,\alpha 2} \; ; \; \vec{\phi}_{,\alpha 22} \; ; \; a_{\alpha\beta} \; ; \; a \; ; \; b_{\alpha\beta} \; ; \; R_C \; ; \; \Gamma^{\beta}_{2\alpha} \; ; \; \Gamma^{2}_{11} \; . \qquad (4.3.25)$$

(ii) *the following functions are discontinuous on symmetry axis* $\xi^1 = 0$:

$$\vec{\phi}_{,11} \; ; \; \vec{\phi}_{,11\alpha} \; ; \; \Gamma^{1}_{11} \; . \qquad (4.3.26)$$

Nevertheless, let us note that

$$\vec{\phi}_{,11} \, , \, \vec{\phi}_{,11\alpha} \in (L^{\infty}(\Omega))^3 \; ; \; \Gamma^{1}_{11} \in L^{\infty}(\Omega) \; . \qquad (4.3.27)$$

Also, let us mention that the continuity in $\bar{\Omega}$ of coefficients $a_{\alpha\beta}$ and $b_{\alpha\beta}$ of both fundamental forms implies the continuity in $\bar{\Omega}$ of *normal curvatures* of the middle surface.

■

4.4. Definition of the arch dam thickness

According to section 4.1, we take the definition of the thickness given by COYNE and BELLIER [1977] by assuming that this one is *measured along the normal* to the middle surface. This definition uses the *arc length* s of the middle line at level ξ^2 = constant. By fixing the origin of the middle line in the symmetry plane of the dam, relation (4.3.16) implies

$$s = \frac{|\xi^1|}{\xi^1} \, \frac{e^{\alpha\theta_\circ |\xi^1|} - 1}{\sin 40^{\circ}} \, \rho_\circ(\xi^2) \qquad (4.4.1)$$

Then, Figure 4.1.3 shows that the *thickness of the dam* at point (ξ^1, ξ^2), i.e.,

$$e : (\xi^1, \xi^2) \in \bar{\Omega} \mapsto e(\xi^1, \xi^2) \qquad (4.4.2)$$

is given by

$$\left.\begin{aligned} e(\xi^1,\xi^2) &= 8 + 0.248Z_\circ\,\xi^2 - 0.000003(Z_\circ\xi^2)^3 \\ &+ 2.10^{-8}\,(Z_\circ\xi^2)^2\,[1 + 0.003Z_\circ\xi^2]\left[\frac{e^{\alpha\theta_\circ|\xi^1|}-1}{\sin 40^\circ}\,\rho_\circ(\xi^2)\right]^2 \end{aligned}\right\} \quad (4.4.3)$$

CHAPTER 5

VARIATIONAL FORMULATION OF THE ARCH DAM PROBLEM

Orientation :

First, we examine how to take into account three kinds of loads which are standard in practical dam problems, i.e., *gravitational*, *hydrostatic* and *thermal* loads. For each, we obtain an expression of the work associated to a displacement $\vec{U}$ of the particles of the dam. Next, we approximate these expressions in order to obtain a formulation set on the middle surface of the dam *which only uses the displacement field* $\vec{u}$ *of the middle surface*. Then we can give a variational formulation of the arch dam problem which enters exactly in the frame of KOITER [1966, 1970]'s model.

5.1. Gravitational loads (due to the weight of the dam)

With the notations of Figure 5.1.1, the loading due to the weight of the dam has a volume density

$$\vec{G}\,dv = \rho_1 g_o\, \overrightarrow{e_3}\, \sqrt{g}\, d\xi^1\, d\xi^2\, d\xi^3 \qquad (5.1.1)$$

where

$$\left.\begin{array}{l} \rho_1 = \text{mass density of the concrete in the undeformed configuration ;} \\ g_o = \text{gravitational acceleration ;} \\ \vec{e}_3 = \text{the vector defined in Figure 4.2.1 ;} \\ \sqrt{g} = (\vec{g}_1 \times \vec{g}_2).\vec{g}_3 \text{ , } \vec{g}_i \text{ defined by (1.2.3). In particular, we have} \\ \quad dV = \sqrt{g}\, d\xi^1\, d\xi^2\, d\xi^3. \end{array}\right\} \qquad (5.1.2)$$

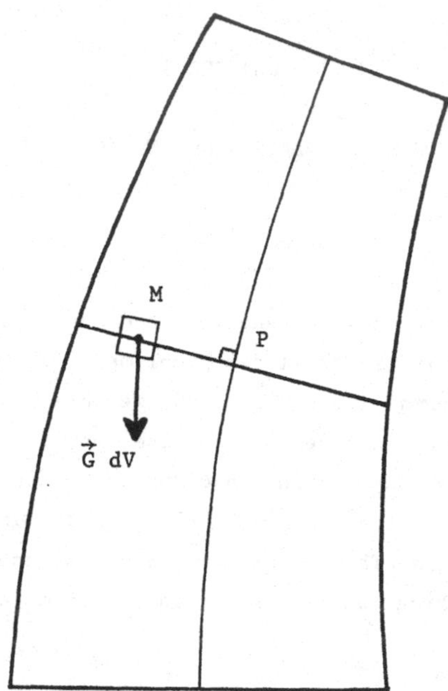

Figure 5.1.1 : Gravitational loading of the dam

Thus, the work of *gravitational loading* of the dam associated to a displacement field $\vec{U}$ of the dam is given by

$$T_G = \int_{\mathcal{C}} \rho_1 g_o \, \vec{e}_3 . \vec{U} \sqrt{g} \, d\xi^1 \, d\xi^2 \, d\xi^3 \quad . \tag{5.1.3}$$

But relation (1.3.20) describes a mapping of the field $\vec{U}$ from the displacement field $\vec{u}$ of the middle surface :

$$\vec{U} = \vec{u} - \xi^3 \, (u_{3|\alpha} + b^\lambda_\alpha u_\lambda) \, \vec{a}^\alpha \quad . \tag{5.1.4}$$

Relations (1.1.15), (1.1.16), (1.2.3) and (5.1.2) give after linearization

$$\sqrt{\frac{g}{a}} = 1 - \xi^3 b^\alpha_\alpha \ . \tag{5.1.5}$$

Hence, expression (5.1.3) becomes after integration with respect to ξ^3 :

$$T_G \simeq \int_\Omega \rho_1 g_o \; e \; \vec{e_3}.\vec{u} \;\; \sqrt{a} \; d\xi^1 \; d\xi^2 \; , \qquad (5.1.6)$$

where e denotes the thickness of the dam defined by (4.4.3). In this approximate expression, we note that the *moments of order* 1 are zero since the integrating function is symmetric with respect to ξ^3. *Moments of order greater or equal to* 2 are neglected since we work on the frame of linear theory of KOITER. But, we have

$$\vec{u} = u_i \vec{a}^i = u_\alpha a^{\alpha\beta} \vec{a}_\beta + u_3 \vec{a}_3 \; ,$$

or, with (4.3.1) and (4.3.2)

$$\vec{u}.\vec{e_3} = (a^{12} u_1 + a^{22} u_2) Z_\circ + (\vec{a_3}.\vec{e_3}) u_3 \; .$$

Hence,

$$T_G \simeq \int_\Omega \rho_1 g_o \; e \; [(a^{12} u_1 + a^{22} u_2) Z_\circ + (\vec{a_3}.\vec{e_3}) u_3] \; \sqrt{a} \; d\xi^1 \; d\xi^2 \qquad (5.1.7)$$

Coefficients a^{12}, a^{22} and a are determined by relations (1.5.13) and (1.5.14). Moreover, relations (1.1.3), (1.1.15) and (1.1.16) imply

$$(\vec{a_3}.\vec{e_3}) \sqrt{a} = (\vec{a_1} \times \vec{a_2}).\vec{e_3} \; ,$$

or, with relations (4.3.1) and (4.3.2) :

$$\left.\begin{aligned} (\vec{a_3}.\vec{e_3}) \sqrt{a} = & - \frac{\theta_\circ e^{\alpha\theta_\circ |\xi^1|} \rho_\circ(\xi^2)}{\cos 40^\circ} \; . \\ & \times \; [\rho'_\circ(\xi^2) \left\{ e^{\alpha\theta_\circ|\xi^1|} \cos 40^\circ - \cos(\theta_\circ|\xi^1| - 40^\circ) \right\} \\ & + \left\{ 0.269 Z_\circ - 0.0000255 (Z_\circ)^3 \; (\xi^2)^2 \right\} \cos(\theta_\circ|\xi^1|)] \end{aligned}\right\} \qquad (5.1.8)$$

5.2. Hydrostatic loads (due to the water pressure)

With notations of Figure 4.2.2, we assume that the dam is filled with water from a given height Z_1 , $0 \le Z_1 \le 157$, i.e.,

$$\left.\begin{array}{ll} 0 \le Z \le Z_1 \Leftrightarrow 0 \le \xi^2 \le \frac{Z_1}{Z_o} & : \text{no water on the upstream wall ;} \\ Z_1 \le Z \le 157 \Leftrightarrow \frac{Z_1}{Z_o} \le \xi^2 \le 1 & : \text{water pressure on the upstream wall.} \end{array}\right\} \quad (5.2.1)$$

According to GERMAIN [1973, p.203], water induces a pressure upon the upstream wall of the dam given by

$$p = \begin{cases} 0 & \text{if } 0 \le \xi^2 \le \frac{Z_1}{Z_o} , \\ \rho_2 g_o Z_o (\xi^2 - \frac{Z_1}{Z_o}) & \text{if } \frac{Z_1}{Z_o} \le \xi^2 \le 1 , \end{cases} \quad (5.2.2)$$

where ρ_2 denotes the *mass density of water*, e.g., the mass per unit volume of water, and where g_o denotes *gravitational acceleration*. It follows that the action of water on the dam is defined by loading surface density $- p\vec{n}$ applied on every area element of dipped part of the upstream wall of the dam. By $\vec{n}$ we denote the *unit normal vector to this upstream wall*, directed to the external part of the dam (see Figure 5.2.1).

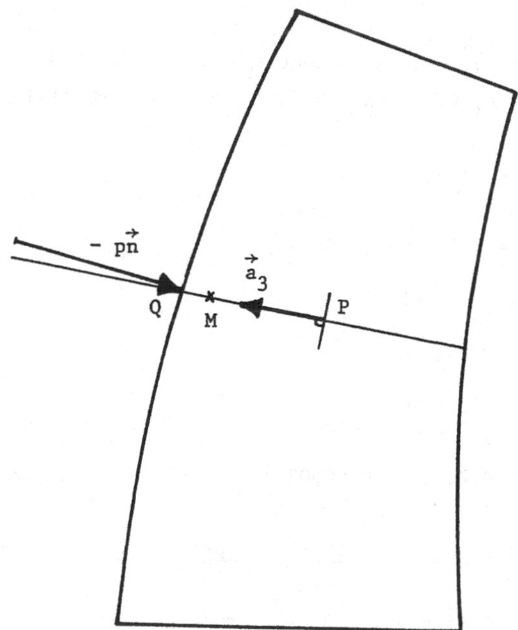

Figure 5.2.1 : Water pressure loading

Let us now find an approximation of the *work* T_H *done by this loading* when associated with a displacement field $\vec{u}$ of the middle surface. According to expression (1.2.2), the upstream wall of the dam is generated by point Q such that

$$\overrightarrow{OQ} = \vec{\phi}(\xi^1,\xi^2) + \frac{1}{2} e(\xi^1,\xi^2)\, \overrightarrow{a_3}(\xi_1,\xi_2) \ , \ (\xi^1,\xi^2) \in \bar{\Omega} \ , \tag{5.2.3}$$

where the mapping $\vec{\phi}$ is determined by relations (4.2.2) (4.2.4). As in the definition of the vector $\overrightarrow{a_3}$, the unit external normal vector to the upstream wall of the dam is given by

$$\vec{n} = \frac{(\overrightarrow{OQ})_{,1} \times (\overrightarrow{OQ})_{,2}}{|(\overrightarrow{OQ})_{,1} \times (\overrightarrow{OQ})_{,2}|} \tag{5.2.4}$$

Relations (1.1.2) and (1.1.14) imply

$$(\overrightarrow{OQ})_{,\alpha} = (\delta^\gamma_\alpha - \frac{1}{2} e\, b^\gamma_\alpha)\, \vec{a}_\gamma + \frac{1}{2} e_{,\alpha}\, \overrightarrow{a_3}$$

so that, with relations (1.1.15) to (1.1.17),

$$\left.\begin{aligned}
&(\overrightarrow{OQ})_{,1} \times (\overrightarrow{OQ})_{,2} \\
&\quad = \sqrt{a}\, \{ -\frac{1}{2} [e_{,1} - \frac{1}{2} e_{,1}\, e\, b^2_2 + \frac{1}{2} e\, e_{,2}\, b^2_1]\, \vec{a}^1 \\
&\qquad - \frac{1}{2} [e_{,2} + \frac{1}{2} e_{,1}\, e\, b^1_2 - \frac{1}{2} e\, e_{,2}\, b^1_1]\, \vec{a}^2 \\
&\qquad + [1 - \frac{1}{2} e\, (b^1_1 + b^2_2) + \frac{1}{4} e^2\, (b^1_1\, b^2_2 - b^1_2\, b^2_1)]\, \vec{a}^3 \}
\end{aligned}\right\}$$

By keeping only linear terms with respect to e or $e_{,\alpha}$, we obtain

$$(\overrightarrow{OQ})_{,1} \times (\overrightarrow{OQ})_{,2} \simeq \sqrt{a}\, \{ -\frac{1}{2} e_{,\alpha}\, \vec{a}^\alpha + (1 - \frac{1}{2} e\, b^\alpha_\alpha)\, \vec{a}^3 \} \tag{5.2.5}$$

The work done by surface loading $- p\vec{n}$ associated to a displacement $\vec{U}(\xi^1,\xi^2, \frac{e}{2})$ of the upstream wall $\mathfrak{J}$ of the dam is given by

$$T_H = \int_{\mathfrak{J}} (- p)\vec{n}.\vec{U}\, d\mathfrak{J} \quad . \tag{5.2.6}$$

With relation (1.3.20), we get

$$\vec{U} = \vec{u} - \frac{e}{2} (u_{3|\alpha} + b^{\lambda}_{\alpha} u_{\lambda}) \vec{a}^{\alpha} \tag{5.2.7}$$

Moreover, as in (1.1.19), area element $d\mathfrak{J}$ of the upstream wall of the dam is given by

$$d\mathfrak{J} = |(\overrightarrow{OQ})_{,1} \times (\overrightarrow{OQ})_{,2}| \, d\xi^1 \, d\xi^2 \quad . \tag{5.2.8}$$

Then, relations (5.2.4) (5.2.5) and (5.2.8) yield

$$\vec{n} \, d\mathfrak{J} \simeq \sqrt{a} \, \{ -\frac{1}{2} e_{,\alpha} \vec{a}^{\alpha} + (1 - \frac{1}{2} e \, b^{\alpha}_{\alpha}) \vec{a}^3 \} \, d\xi^1 \, d\xi^2 \tag{5.2.9}$$

By substituting relations (5.2.7) and (5.2.9) into relation (5.2.6), we obtain the following expression (linearized with respect to e or $e_{,\alpha}$) of the *work* T_H *done by hydrostatic loading* :

$$T_H \simeq \int_{\Omega} p \, [\frac{1}{2} e_{,\beta} \, u_{\alpha} \, a^{\alpha\beta} - (1 - \frac{1}{2} e \, b^{\beta}_{\beta}) u_3] \sqrt{a} \, d\xi^1 \, d\xi^2 . \tag{5.2.10}$$

Moreover, let us note that expression (4.4.3) implies :

$$\left. \begin{aligned} e_{,1}(\xi^1,\xi^2) &= 4.10^{-8} \, \theta_o (Z_o)^2 \, (\xi^2)^2 \, (\rho_o(\xi^2))^2 \, \frac{e^{\alpha\theta_o|\xi^1|}}{\sin 40^{\circ} \, \cos 40^{\circ}} \\ &\times \frac{|\xi^1|}{\xi^1} \, [1 + 0.003 Z_o \xi^2] [e^{\alpha\theta_o|\xi^1|} - 1] \, , \end{aligned} \right\} \tag{5.2.11}$$

$$\left. \begin{aligned} e_{,2}(\xi^1,\xi^2) &= 0.248 Z_o - 0.000009 (Z_o)^3 \, (\xi^2)^2 \\ &+ 2.10^{-8} \, (Z_o)^2 \left(\frac{e^{\alpha\theta_o|\xi^1|} - 1}{\sin 40^{\circ}} \right)^2 \xi^2 \, \rho_o(\xi^2) \\ &\times [(2 + 0.009 Z_o \xi^2) \, \rho_o(\xi^2) + 2\xi^2 \, (1 + 0.003 Z_o \xi^2) \rho_o'(\xi^2)] . \end{aligned} \right\} \tag{5.2.12}$$

5.3. Thermal loads

A change of temperature in an elastic continuous medium induces deformations due to *thermal dilatation* (see for instance LANDAU and LIFCHITZ [1967, §6]). In what follows, we shall take into account these thermal effects using the simplest stationnary form. To obtain the corresponding equations we proceed in two steps : (i) *we record the equations of thermoelasticity* for a definition of the dam using a system of curvilinear coordinates (ξ^1,ξ^2,ξ^3) ; (ii) from these equations, *we derive approximate bidimensional equations* set on the planar reference domain Ω.

Step 1 : Thermoelasticity equations

For a thorough study of thermoelasticity equations we refer to CARLSON [1972]. Also, we have used GREEN and ADKINS [1970] who give the equations for a description of the continuous medium with the help of a system of *general curvilinear coordinates*.

In its reference configuration, the dam is assumed to be at rest and at a *uniform temperature* $\theta_\circ$. Under the actions of loads described in sections 5.1 and 5.2 and under the action of surface thermal loading (we detail farther), the dam takes deformations. The thermoelasticity equations given below allow us to find : (i) *the displacement field* $\vec{U}$ which associates with any particle of the undeformed configuration its position in the deformed configuration ; (ii) *the temperature* $\theta\ (\xi^1,\xi^2,\xi^3)$ of the same particle.

We denote respectively by σ^{*ij}, γ^*_{ij} the *contravariant* and *covariant components of stress* and *strain tensors* of the dam. According to GREEN and ADKINS [1970, (8.7.6)], *stress-strain-temperature relations* are given by

$$\sigma^{*ij} = E^{*ijk\ell}\ \gamma^*_{k\ell} - \frac{E\bar{\alpha}}{1-2\nu}(\theta - \theta_\circ)\ g^{ij} , \tag{5.3.1}$$

where

$$\left.\begin{aligned} &E^{*ijk\ell} = \frac{E}{2(1+\nu)}\ [g^{ik}\ g^{j\ell} + g^{i\ell}\ g^{jk} + \frac{2\nu}{1-2\nu}\ g^{ij}\ g^{k\ell}] \\ &E = \text{YOUNG's modulus},\ \nu = \text{POISSON's coefficient} \end{aligned}\right\} \tag{5.3.2}$$

$$\bar{\alpha} = \text{coefficient of linear expansion} \tag{5.3.3}$$

Strain energy of the arch dam :

The *strain energy of the arch dam* associated with a displacement field $\vec{U}$ and with a change of temperature field $\theta - \theta_\circ$ is given by

$$E_d = \frac{1}{2} \int_{\mathcal{C}} \sigma^{*ij} \gamma^*_{ij} \, dV \tag{5.3.4}$$

or equivalently, with (5.3.1)

$$E_d = E_{d1} + E_{d2} \tag{5.3.5}$$

where

$$E_{d1} = \frac{1}{2} \int_{\mathcal{C}} E^{*ijk\ell} \gamma^*_{k\ell} \gamma^*_{ij} \, dV \ , \tag{5.3.6}$$

$$E_{d2} = \frac{-E\bar{\alpha}}{1-2\nu} \int_{\mathcal{C}} (\theta - \theta_\circ) \, g^{ij} \gamma^*_{ij} \, dV \ . \tag{5.3.7}$$

In section 1.3, we have seen that a good approximation of E_{d1} is given by

$$E_{d1} \simeq \frac{1}{2} a(\vec{u},\vec{u}) \ . \tag{5.3.8}$$

Here, $\vec{u}$ denotes the displacement field of the middle surface and $a(\vec{u},\vec{u})$ is defined by (1.3.26).

Now, let us consider *thermal component* E_{d2} *of the strain energy* E_d. Thermal loads arise from changes of temperature applied on the upstream and downstream walls of the dam. The temperature field $\theta(\xi^1,\xi^2,\xi^3)$ inside of the dam is governed by the equation

$$\text{div}\,(k \text{ grad } \theta) = 0 \ ,$$

where k is *thermal conductivity coefficient* of the material. In this work, we assume that k is constant so that the previous equation becomes

$$\Delta\theta = 0 \ . \tag{5.3.9}$$

<u>Step 2</u> : <u>Approximate bidimensional equations</u>

Since we work within the framework of the linear theory, we assume from now on that the following hypothesis is satisfied :

<u>Hypothesis 5.3.1</u> : <u>The temperature field θ varies linearly through the thickness of the dam.</u>

■

Let us denote by T the *change of temperature field*, i.e.,

$$T(\xi^1,\xi^2,\xi^3) = (\theta - \theta_\circ)(\xi^1,\xi^2,\xi^3) \tag{5.3.10}$$

Since the field $\theta_\circ$ is assumed to be constant, the field T satisfies also Hypothesis 5.3.1. Let

$$\left.\begin{aligned} T_{up}(\xi^1,\xi^2) &= T(\xi^1,\xi^2,\tfrac{e}{2}) \ , \\ T_{do}(\xi^1,\xi^2) &= T(\xi^1,\xi^2,-\tfrac{e}{2}) \ , \end{aligned}\right\} \tag{5.3.11}$$

be the *change of temperature fields of upstream and downstream walls of the dam*, respectively. In the following, these fields T_{up} and T_{do} will be the data of the problem. Then, Hypothesis 5.3.1 leads immediately to

$$T(\xi^1,\xi^2,\xi^3) = T_1(\xi^1,\xi^2) + \xi^3 T_2(\xi^1,\xi^2) \tag{5.3.12}$$

with

$$T_1(\xi^1,\xi^2) = \frac{1}{2}\left(T_{up}(\xi^1,\xi^2) + T_{do}(\xi^1,\xi^2)\right) \tag{5.3.13}$$

$$T_2(\xi^1,\xi^2) = \frac{1}{e(\xi^1,\xi^2)}\left(T_{up}(\xi^1,\xi^2) - T_{do}(\xi^1,\xi^2)\right) \tag{5.3.14}$$

In expression (5.3.7), it remains to define the linear part of the formulation with respect to ξ^3 of functions g^{ij}, γ^*_{ij} and dV . First, developments (1.3.7) and the relation $[g^{ij}] = [g_{ij}]^{-1}$ lead to the following approximations :

$$\left.\begin{aligned} g^{\alpha\beta} &= a^{\alpha\beta} + 2b^{\alpha\beta}\,\xi^3 \ , \\ g^{\alpha 3} &= g^{3\alpha} = 0 \ , \\ g^{33} &= 1 \ . \end{aligned}\right\} \tag{5.3.15}$$

Moreover, relations (1.3.8) to (1.3.11) give

$$\left.\begin{aligned} \gamma^*_{\alpha\beta} &= \gamma_{\alpha\beta} - \xi^3\bar{\rho}_{\alpha\beta} \,, \\ \gamma^*_{\alpha 3} &= \gamma^*_{3\alpha} = 0 \,, \\ \gamma^*_{33} &= - \frac{\nu}{1-\nu} [\gamma^\alpha_\alpha + \xi^3(2b^\alpha_\eta \, \gamma^\eta_\alpha - \bar{\rho}^\alpha_\alpha)] \quad . \end{aligned}\right\} \tag{5.3.16}$$

It follows that

$$g^{ij} \, \gamma^*_{ij} = \frac{1-2\nu}{1-\nu} [\gamma^\alpha_\alpha + \xi^3(2b^\alpha_\eta \, \gamma^\eta_\alpha - \bar{\rho}^\alpha_\alpha)] \,. \tag{5.3.17}$$

Finally, the *volume element* dV is given by

$$dV = \sqrt{g} \;\; d\xi^1 \, d\xi^2 \, d\xi^3 \,,$$

or, with (5.1.5) :

$$dV = (1 - \xi^3 \, b^\alpha_\alpha) \sqrt{a} \, d\xi^1 \, d\xi^2 \, d\xi^3 \quad . \tag{5.3.18}$$

By substituting the expressions (5.3.10), (5.3.12), (5.3.17) and (5.3.18) into expression (5.3.7), we obtain

$$\left.\begin{aligned} E_{d2} = - \frac{1}{2} \frac{E\bar{\alpha}}{1-\nu} \int_{\mathcal{C}} &[T_1 + \xi^3 T_2][\gamma^\alpha_\alpha + \xi^3(2b^\alpha_\eta \, \gamma^\eta_\alpha - b^\alpha_\alpha \, \gamma^\eta_\eta - \bar{\rho}^\alpha_\alpha)] \\ &\sqrt{a} \, d\xi^1 \, d\xi^2 \, d\xi^3 \end{aligned}\right\} \tag{5.3.19}$$

from which, we deduce

$$\left.\begin{aligned} E_{d2} = - \frac{1}{2} \frac{E\bar{\alpha}}{1-\nu} \int_\Omega &[e \; T_1 \, \gamma^\alpha_\alpha + \frac{e^3}{12} T_2(2b^\alpha_\eta \, \gamma^\eta_\alpha - b^\alpha_\alpha \, \gamma^\eta_\eta - \bar{\rho}^\alpha_\alpha)] \\ &\sqrt{a} \, d\xi^1 \, d\xi^2 \quad . \end{aligned}\right\} \tag{5.3.20}$$

Here the thickness e and temperature resultants T_1 and T_2 are respectively given by relations (4.4.3), (5.3.13) and (5.3.14).

<u>Remark 5.3.1</u> :

In agreement with previous linearizations, we have neglected term in $(\xi^3)^2$ in the second factor of the integrant function of the expression (5.3.19). Such a term would lead to contributions of order 3 with respect to the thickness. On the contrary, taking into account the definition (5.3.14), we keep the term $(\xi^3)^2 \; T_2(2b^\alpha_\eta\gamma^\eta_\alpha - b^\alpha_\alpha\gamma^\eta_\eta - \bar{\rho}^\alpha_\alpha)$ which leads to contribution of order 2 with respect to the thickness in the expression (5.3.20). Moreover, this term is essential in following situation :

$$T_{am}(\xi^1,\xi^2) = - T_{av}(\xi^1,\xi^2) \quad ,$$

for which we have

$$\left.\begin{aligned} T_1(\xi^1,\xi^2) &= 0 \\ T_2(\xi^1,\xi^2) &= \frac{2\ Tam(\xi^1,\xi^2)}{e(\xi^1,\xi^2)} \quad . \end{aligned}\right\}$$

This situation can arise if we notice that it corresponds, for instance, to a reheating of downstream wall and to a growing cold of upstream wall of same amplitude. ∎

<u>Remark 5.3.2</u> :

The approximation (5.3.12) of change of temperature is somewhat crude, but is seems sufficient in present case. For a thorough study of thermoelasticity applied to thin shells, we refer to GREEN and NAGHDI [1970, 1978] or NAGHDI [1972]. An example taking into account thermal effects for a spherical shell is given by CRAINE [1968]. ∎

<u>Remark 5.3.3</u> :

The previous analysis does not take into account possible changes of temperature at the top of the dam. If such were the case, we must add a boundary term at the second member which gives the expression of thermal work done by the corresponding force resultant and moment resultant of thermal loads on the crest middle line of the dam. ∎

5.4. Variational formulation of the arch dam problem

By combining results of sections 5.1 to 5.3, we obtain the *energy* $J(\vec{v})$ *of the dam* associated to a displacement field $\vec{v}$ of the middle surface. Relations (1.3.25), (5.1.7), (5.2.10), (5.3.8) and (5.3.20) lead to the expression

$$J(\vec{v}) = \frac{1}{2}\, a(\vec{v},\vec{v}) - f(\vec{v}) \tag{5.4.1}$$

with

$$\left.\begin{aligned} f(\vec{v}) = {} & \frac{E\bar{\alpha}}{2(1-\nu)} \int_\Omega [e\, T_1\, \gamma^\beta_\beta + \frac{e^3}{12}\, T_2 (2b^\beta_\eta\, \gamma^\eta_\beta - b^\beta_\beta\, \gamma^\eta_\eta - \bar{\rho}^\beta_\beta)] \sqrt{a}\, d\xi^1\, d\xi^2 \\ & + \int_\Omega p\, [\frac{1}{2}\, e_{,\beta}\, v_\eta\, a^{\eta\beta} - (1 - \frac{1}{2}\, e\, b^\beta_\beta)\, v_3] \sqrt{a}\, d\xi^1\, d\xi^2 \\ & + \int_\Omega \rho_1 g_o\, e\, [(a^{12}\, v_1 + a^{22}\, v_2)\, Z_o + (\overrightarrow{a_3}\cdot\overrightarrow{e_3})\, v_3] \sqrt{a}\, d\xi^1\, d\xi^2 \, . \end{aligned}\right\} \tag{5.4.2}$$

According to BERNADOU and CIARLET [1976], the *space* $\vec{V}$ *of admissible displacements* is defined by

$$\left.\begin{aligned} \vec{V} = \{\vec{v} \mid & \vec{v} \in (H^1(\Omega))^2 \times H^2(\Omega)\ ,\ \vec{v}|_{\Gamma_o} = \vec{0}\ , \\ & \frac{\partial v_3}{\partial n}\Big|_{\Gamma_o} = 0\} \end{aligned}\right\} \tag{5.4.3}$$

where, in the present case, Γ_o denotes the part of the boundary Γ of domain Ω which is associated with lateral and lower (Z = 157) borders of the middle surface of the dam. Part $\Gamma_1 = \Gamma - \Gamma_o$ associated to upper border (Z = 0) of the middle surface of the dam is *free*.

Moreover, let us assume that

$$T_\alpha \in L^2(\Omega)\quad , \alpha = 1,2\ . \tag{5.4.4}$$

Properties (4.3.27) and the continuity of functions $a^{\alpha\beta}$, b^α_β imply that

$$\left.\begin{aligned} \gamma^\alpha_\beta(\vec{v}) \in L^2(\Omega) \quad & , \quad \forall \vec{v} \in \vec{V} \\ \bar{\rho}^\alpha_\beta(\vec{v}) \in L^2(\Omega) \quad & , \quad \forall \vec{v} \in \vec{V} \end{aligned}\right\} \tag{5.4.5}$$

where components γ^α_β and $\bar{\rho}^\alpha_\beta$ are defined by relations (1.3.21), (1.3.22) and (1.3.25). It is easy to check that expression (5.4.1) makes sense for any $\vec{v} \in \vec{V}$ and for any $T_\alpha \in L^2(\Omega)$, $\alpha = 1,2$.

The symmetry of the bilinear form a(.,.) defined in (1.3.26) makes it possible to construct the following equivalent formulations of the problem :

Variational formulation :

For $T_\alpha \in L^2(\Omega)$, $\alpha = 1,2$, find $\vec{u} \in \vec{V}$ such that

$$a(\vec{u},\vec{v}) = f(\vec{v}) \ , \quad \forall \vec{v} \in \vec{V} \ . \tag{5.4.6}$$

■

Minimization formulation :

For $T_\alpha \in L^2(\Omega)$, $\alpha = 1,2$, find $\vec{u} \in \vec{V}$ minimizing the functional

$$J : \vec{v} \in \vec{V} \to J(\vec{v}) = \frac{1}{2} a(\vec{v},\vec{v}) - f(\vec{v}) \quad . \tag{5.4.7}$$

■

Then, we have following *existence and uniqueness theorem :*

Theorem 5.4.1 : *Problem* (5.4.6) *has one and only one solution* $\vec{u} \in \vec{V}$. *This function* $\vec{u}$ *is also the unique solution of problem* (5.4.7).

Proof :

The proof is similar to that of BERNADOU and CIARLET [1976]. In that work, it is assumed for simplicity that

$$\vec{\phi} \in (\mathcal{C}^3(\bar{\Omega}))^3 \quad , e = \text{constant.}$$

The reader can check that properties stated in Remark 4.3.1 allow us to extend the existence and uniqueness result to present case.

■

5.5. Another expression for the linear form f(.)

Similarly to Theorem 1.5.2, we prove :

Theorem 5.5.1 : *The linear form* f(.) *defined in* (5.4.2) *can be written*

$$f(\vec{v}) = \int_\Omega {}^t\mathbf{F}\mathbf{V} d\xi^1 \, d\xi^2 \quad , \tag{5.5.1}$$

where the column matrix **V** *is given in* (1.5.2) *and where the column matrix* **F** *is given by*

$$
\left.\begin{aligned}
{}^t\mathbf{F} &= \frac{E\bar{\alpha}}{2(1-\nu)}\, e \sqrt{a}\ [T_1\, \Lambda^\beta_\beta + \frac{e^2}{12}.T_2\ (2b^\beta_\eta\, \Lambda^\eta_\beta - b^\beta_\beta\, \Lambda^\eta_\eta - M^\beta_\beta)] \\
&+ [\frac{p}{2} \sqrt{a}\ e_{,\beta}\ a^{\beta 1} + \rho_1 g_o\ e \sqrt{a}\ Z_o\ a^{12}\ ;\ 0\ ;\ 0\ ; \\
&\frac{p}{2} \sqrt{a}\ e_{,\beta}\ a^{\beta 2} + \rho_1 g_o\ e \sqrt{a}\ Z_o\ a^{22}\ ;\ 0\ ;\ 0\ ; \\
&- p \sqrt{a}(1 - \frac{1}{2}\, e\, b^\beta_\beta) + \rho_1 g_o\ e \sqrt{a}\ (\vec{a_3}.\vec{e_3})\ ;\ 0\ ;\ 0\ ;\ 0\ ;\ 0\ ;\ 0]\ .
\end{aligned}\right\}\ (5.5.2
$$

The matrices Λ^η_β *and* M^η_β *are defined in* (1.5.6) *and* (1.5.9).

∎

CHAPTER 6

IMPLEMENTATION - PRESENTATION OF RESULTS

Orientation :

This chapter describes the final step of the implementation : first, we give the values of physical constants and the method of triangulation ; next, we indicate how to take into account the boundary conditions and the symmetry conditions. Then, the linear system so obtained is solved by a CHOLESKI method using the sky-line bandwidth factorization.

From this system, we deduce an approximation of the displacement field and of the stress field at any point of the dam. We conclude by giving explicitly physical components of the stresses.

6.1. Values of physical constants

For convenience, we gather in this section the values of physical constants that we have used for calculations.

YOUNG's modulus and POISSON's coefficient (see (1.3.24)) :

$$(1.5)10^4 \leq E \leq (3.5)10^4 \text{ M.New/m}^2 \;,\; \nu = 0.2 \;.$$

For computations, we have taken $E = 2\cdot 10^4$ M.New/m^2.

Mass densities of concrete (see (5.1.2)) and of water (see (5.2.2)) :

$$\rho_1 = 2500 \text{ Kg/m}^3 \quad , \quad \rho_2 = 1000 \text{ Kg/m}^3 \quad .$$

Gravitational acceleration (see (5.1.2) and (5.2.2)) :

$$g_\circ = 9.81 \text{ m/s}^2$$

Coefficient of linear expansion of the concrete (see (5.3.3)) :

$$8\ 10^{-6} \leq \bar{\alpha} \leq 12\ 10^{-6} \text{ by Celsius degrees.}$$

For calculations, we have taken $\bar{\alpha} = 10\ 10^{-6}$ by Celsius degree.

6.2. Triangulation

We adopt a regular triangulation of domain Ω shown in Figure 4.2.2. In this aim, we subdivide the intervals

$x = \xi^1 \in [0,1]$ in M equal parts,

$y = \xi^2 \in [0,1]$ in N equal parts,

and next, we subdivide opposite sides in the same ratio. So doing, we obtain a mesh of the domain by using quadrilaterals. Then every quadrilateral is subdivided into two triangles as follows : we start by a triangulation of the lower right quadrilateral by using the diagonal containing the boundary corner. The reasons of this choice will be given

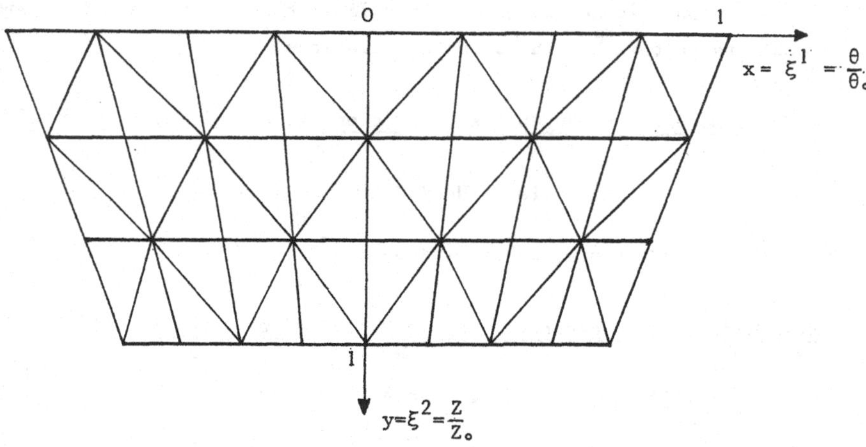

Figure 6.2.1 : Triangulation of the domain Ω (M = 4 , N = 3).

in Remark 6.3.1 below. Next, on quadrilaterals adjacent to the quadrilateral just partitioned, we alternate the diagonal of the triangulation as shown in the Figure 6.2.1, for which M = 4 , N = 3. In this case we denote the triangulation by $\mathscr{C}_{43}$ and, in the general case, by $\mathscr{C}_{MN}$.

6.3. How to take into account boundary conditions

Boundary conditions are of two kinds :

(1) *Clamped conditions* along $\Gamma_\circ$, i.e., along middle lines of lateral borders as well as on the bottom of the dam. According to (2.1.4) (2.1.5), we have to prescribe restrictions of functions $\vec{v_h} \in X_{h1} \times X_{h1} \times X_{h2}$ so that

$$v_{hi}|_{\Gamma_\circ} = 0 \ , \ i = 1,2,3 \ ; \qquad (6.3.1)$$

$$\frac{\partial v_{h3}}{\partial n}|_{\Gamma_\circ} = 0 \ . \qquad (6.3.2)$$

(2) *Free conditions* on $\Gamma_1 = \Gamma - \Gamma_\circ$, i.e., along the crest middle line of the dam. Thus, along Γ_1 we do not have any restrictions to prescribe to functions $\vec{v_h}$.

These two types of boundary conditions are illustrated in Figure 6.3.1.

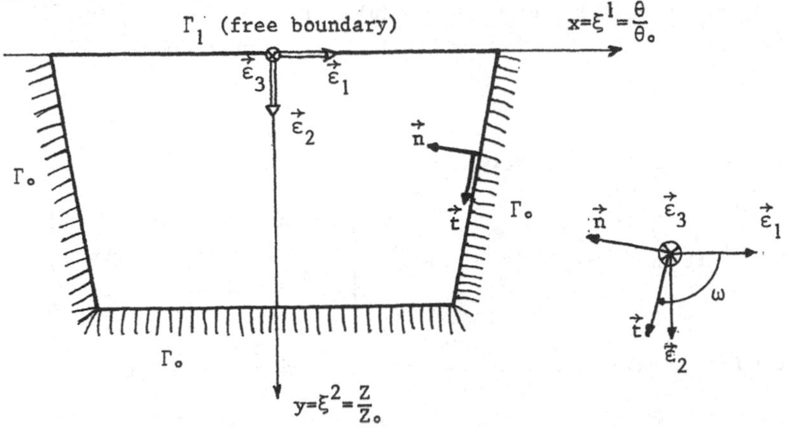

Figure 6.3.1 : Boundary conditions

To take into account boundary conditions of type (6.3.1) or (6.3.2), it is convenient to introduce a new set of degrees of freedom - *the set of border degrees of freedom* - defined at every vertex of triangles located on $\Gamma_\circ$. At these vertices, we substitute this new set for the set of global degrees of freedom which has been defined for instance in (3.1.20). Practically, for a vertex "a" located on $\Gamma_\circ$ we shall distinguish two cases according to

(i) *the boundary* $\Gamma_\circ$ *admits at point* a *only one direction of tangent.* Let us denote by $\vec{t}$ one of the unit tangent vectors. With notations of relation (3.1.2) and Figure 6.3.1, let $\vec{n}$ be the unit vector such that $\vec{n} = \vec{\varepsilon}_3 \times \vec{t}$, $\vec{\varepsilon}_3 = \vec{\varepsilon}_1 \times \vec{\varepsilon}_2$ and let $\omega \in [-\Pi, \Pi[$ be the angle measured along $\vec{\varepsilon}_3$ such that

$$\vec{t} = \cos\omega\, \vec{\varepsilon}_1 + \sin\omega\, \vec{\varepsilon}_2 \,, \tag{6.3.3}$$

$$\vec{n} = -\sin\omega\, \vec{\varepsilon}_1 + \cos\omega\, \vec{\varepsilon}_2 \,. \tag{6.3.4}$$

Then, for all vertices of triangle $a \in \Gamma_\circ$ of this kind, we shall substitute the new set of border degrees of freedom

$$\partial_t v_h(a) \quad , \partial_n v_h(a)$$

for the usual set of global degrees of freedom $\partial_x v_h(a)$ and $\partial_y v_h(a)$. Between both sets, we have the relations :

$$[\partial_x v_h(a) \quad \partial_y v_h(a)] = [\partial_t v_h(a) \quad \partial_n v_h(a)] \begin{bmatrix} \cos\omega & \sin\omega \\ -\sin\omega & \cos\omega \end{bmatrix} \tag{6.3.5}$$

$$[\partial_{xx} v_h(a) \ \partial_{xy} v_h(a) \ \partial_{yy} v_h(a)] =$$

$$= [\partial_{tt} v_h(a) \ \partial_{tn} v_h(a) \ \partial_{nn} v_h(a)] \begin{bmatrix} \cos^2\omega & \cos\omega\sin\omega & \sin^2\omega \\ -\sin 2\omega & \cos 2\omega & \sin 2\omega \\ \sin^2\omega & -\sin\omega\cos\omega & \cos^2\omega \end{bmatrix} \tag{6.3.6}$$

(ii) *the boundary* $\Gamma_\circ$ *admits two independent half-tangents* $\vec{t_1}$ *and* $\vec{t_2}$, *i.e., a is a salient point.*

Let $\overrightarrow{t_1}$, $\overrightarrow{t_2}$ be two unit vectors carried by these half-tangents. By using construction (6.3.3) (6.3.4), we define angles ω_1 and ω_2 such that

$$\left.\begin{aligned} \overrightarrow{t_1} &= \cos\omega_1\ \overrightarrow{e_x} + \sin\omega_1\ \overrightarrow{e_y}\ , \quad \omega_1 \in [-\Pi,\Pi[\ , \\ \overrightarrow{t_2} &= \cos\omega_2\ \overrightarrow{e_x} + \sin\omega_2\ \overrightarrow{e_y}\ , \quad \omega_2 \in [-\Pi,\Pi[\ . \end{aligned}\right\} \tag{6.3.7}$$

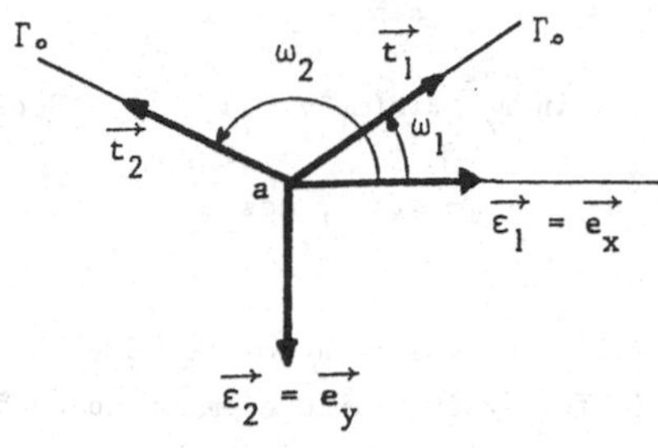

Figure 6.3.2 : Case of a salient point

In Figure 6.3.2, note that the angles ω_1 and ω_2 satisfy the inequalities $-\Pi < \omega_2 < \omega_1 < 0$, due to the orientation of $\overrightarrow{\varepsilon_3} = \overrightarrow{\varepsilon_1} \times \overrightarrow{\varepsilon_2}$.

In this case, we substitute the new set of borders degrees of freedom

$$\partial_{t_1} v_h(a)\ , \quad \partial_{t_2} v_h(a)$$

for the usual set of global degrees of freedom $\partial_x v_h(a)$ and $\partial_y v_h(a)$. Between both sets, we have the relations

$$\left.\begin{aligned} &[\partial_x v_h(a)\ \partial_y v_h(a)] = \\ &\quad = \frac{1}{\sin(\omega_2-\omega_1)}\ [\partial_{t_1} v_h(a)\ \partial_{t_2} v_h(a)] \begin{bmatrix} \sin\omega_2 & -\cos\omega_2 \\ -\sin\omega_1 & \cos\omega_1 \end{bmatrix} \end{aligned}\right\} \tag{6.3.8}$$

(let us point out that a = salient point implies $\omega_2 \neq \omega_1 + k\Pi$, $k = -1, 0, 1$).

Similarly, we get :

$$\left.\begin{aligned}
&[\partial_{xx} v_h(a) \quad \partial_{xy} v_h(a) \quad \partial_{yy} v_h(a)] \\
&= \frac{1}{\sin^2(\omega_2-\omega_1)} [\partial_{t_1t_1} v_h(a) \quad \partial_{t_1t_2} v_h(a) \quad \partial_{t_2t_2} v_h(a)] \times \\
&\times \begin{bmatrix} \sin^2 \omega_2 & - \sin \omega_2 \cos \omega_2 & \cos^2 \omega_2 \\ - 2\sin \omega_1 \sin \omega_2 & \sin(\omega_1 + \omega_2) & - 2\cos \omega_1 \cos \omega_2 \\ \sin^2 \omega_1 & - \sin \omega_1 \cos \omega_1 & \cos^2 \omega_1 \end{bmatrix}
\end{aligned}\right\} \quad (6.3.9)$$

In Figures 6.3.3 to 6.3.5, we indicate how to take into account clamped conditions along Γ_0 for every finite element space introduced in sections 3.2 to 3.6. Let us record that these spaces are constructed with the help of finite elements defined in Figures 2.1.1 to 2.1.5.

The results which appear in Figures 6.3.3 to 6.3.5 are relative to *border degrees of freedom* defined on part $x > 0$ of boundary Γ_0. One could complete by symmetry along the part $x < 0$ of boundary Γ_0. Finally, at the border vertex ($x = 0$, $y = 1$), we need to possibly add symmetry conditions : for these, we return to section 6.4, particularly to Figures 6.4.2 to 6.4.4.

In Figures 6.3.3 to 6.3.5, we adopt the following notations

[0] : the corresponding degree of freedom is zero ;

[1] : the corresponding degree of freedom is really an unknown of the problem.

Practically, we adopt a global numbering scheme for the degrees of freedom for which the degrees of freedom corresponding to prescribed (generalized) forces are numbered consecutively first and the remaining degrees of freedom corresponding to fixed (zero) degrees of freedom are

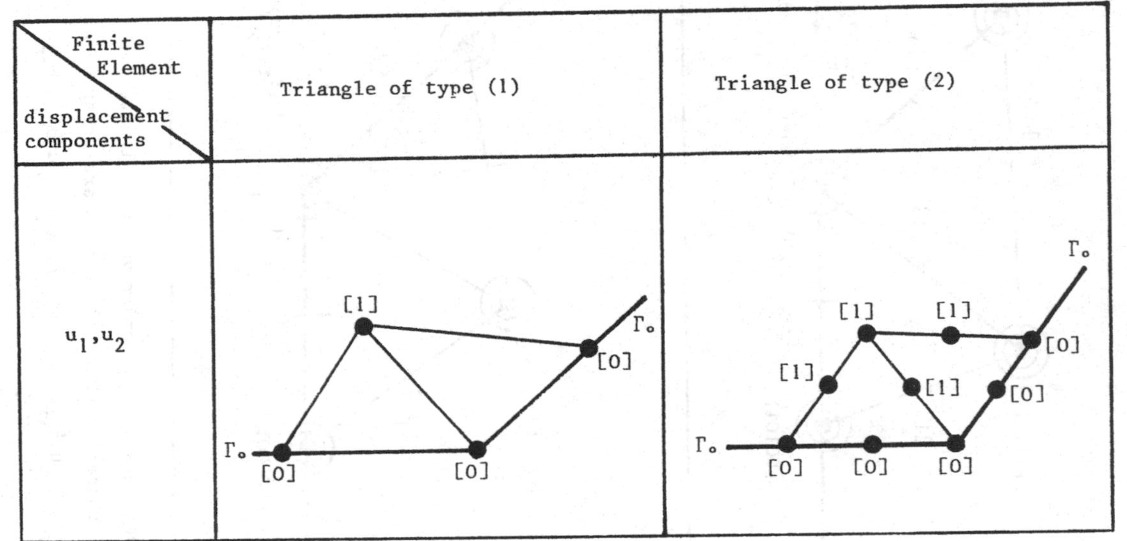

Figure 6.3.3 : Taking into account clamped conditions for triangles of type (1) or (2)

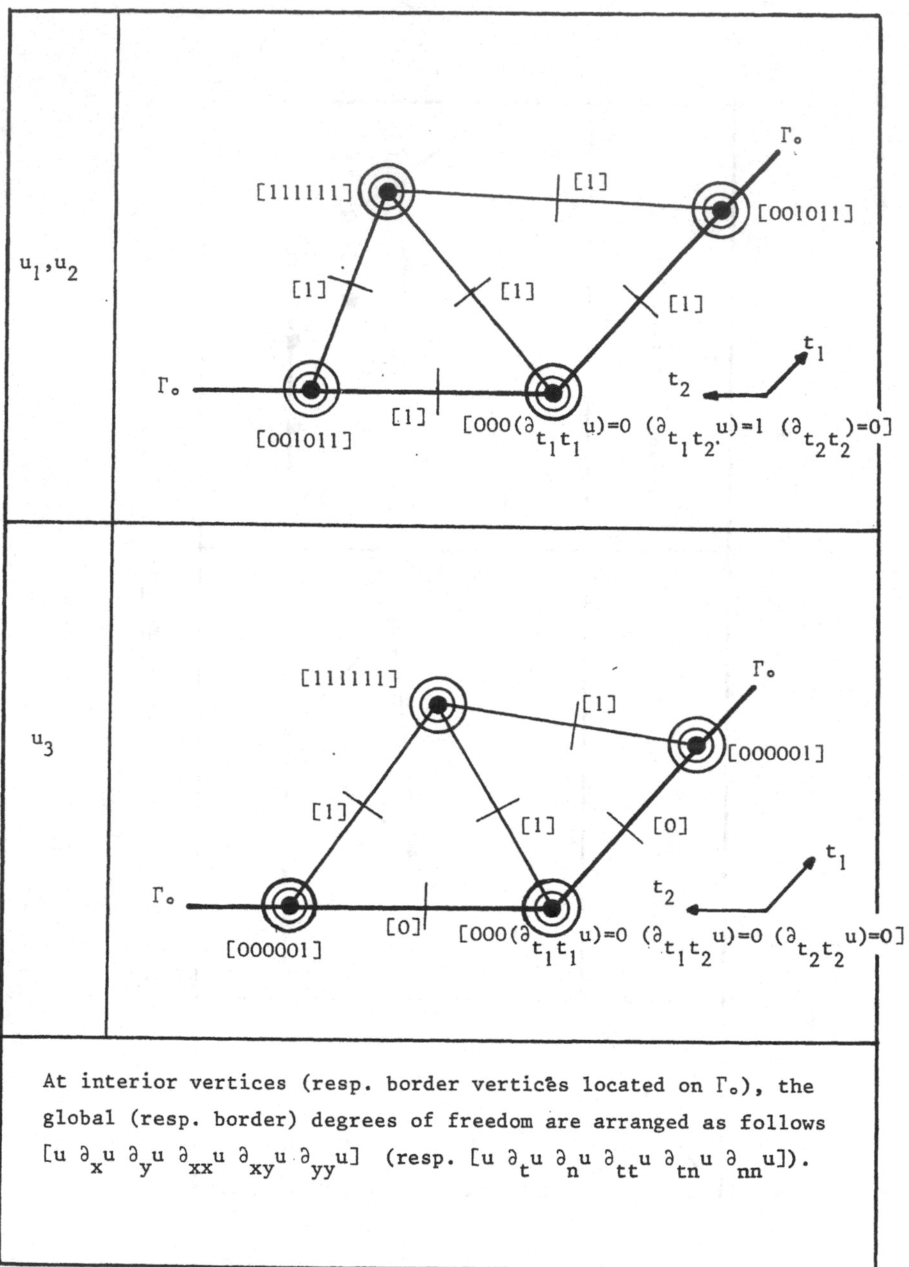

Figure 6.3.4 : Taking into account clamped conditions for the ARGYRIS triangles.

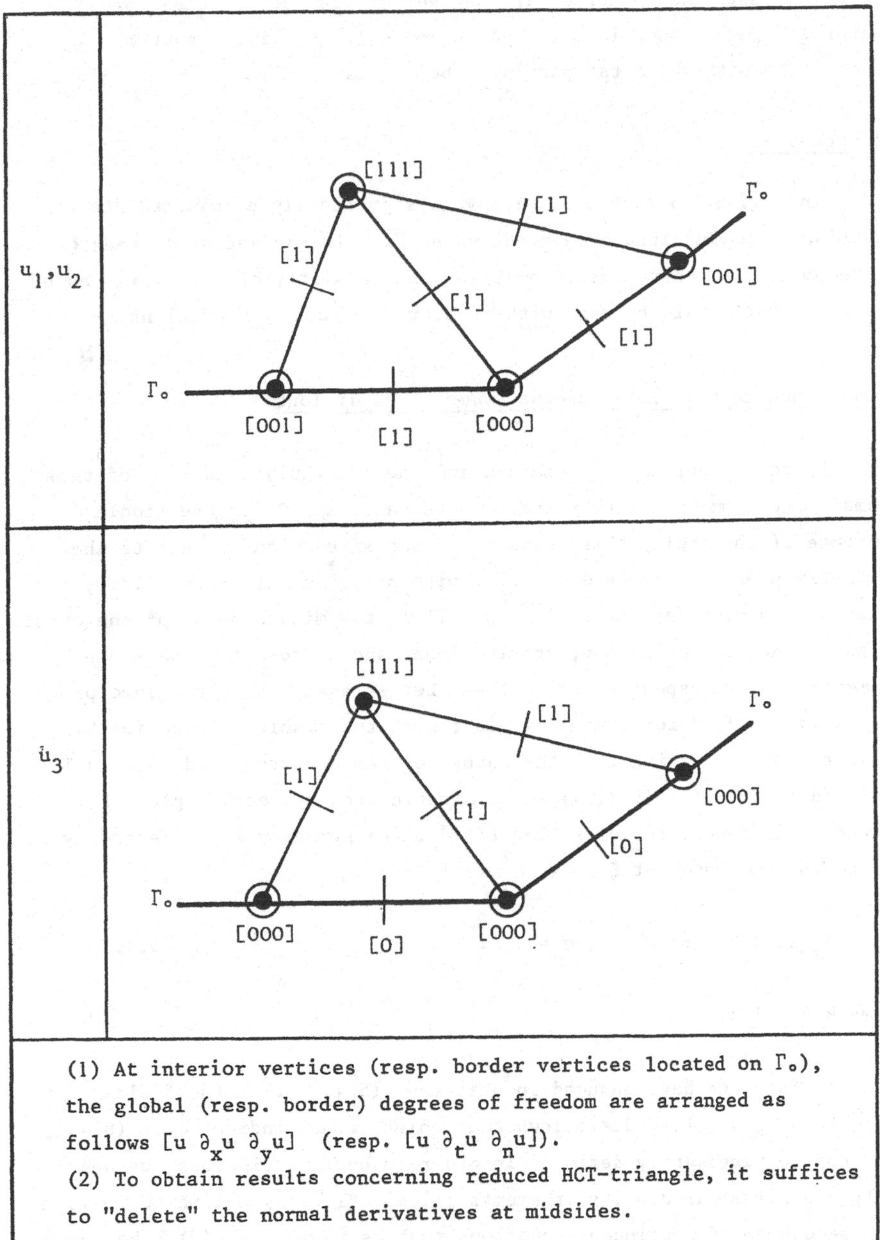

(1) At interior vertices (resp. border vertices located on $\Gamma_\circ$), the global (resp. border) degrees of freedom are arranged as follows $[u\ \partial_x u\ \partial_y u]$ (resp. $[u\ \partial_t u\ \partial_n u]$).

(2) To obtain results concerning reduced HCT-triangle, it suffices to "delete" the normal derivatives at midsides.

Figure 6.3.5 : Taking into account clamped conditions for the complete or reduced HCT-triangles.

numbered last. The vector of all degrees of freedom is, therefore, naturally partitioned into two parts and only the linear system associated with the first part need be solved.

Remark 6.3.1 :

In Fig. 6.3.3 to 6.3.5, we have systematically eliminated the case of a triangle having its three vertices on $\Gamma_\circ$: in the set of degrees of freedom we would have most, sometimes the totality, of these degrees of freedom which would be zero with respect to boundary conditions.

∎

6.4. How to take into account symmetry conditions

By construction, the domain Ω and the triangulations $\mathcal{T}_{MN}$ of this domain are symmetric with respect to the axis x = 0. On the middle surface of the shell, this property is satisfied with respect to the symmetry plane P_S of the dam, i.e., with notations of Figure 4.2.1, the plane containing vectors $\vec{e}_1$, $\vec{e}_3$ as well as the middle point of the crest middle line. By definition, gravitational and hydrostatic loads are symmetric with respect to this plane. Let us assume that the same property is verified for thermal loads ; with the results of section 5.3, this amounts to saying that the change of temperature fields T_1 and T_2 defined in (5.3.13) (5.3.14) are symmetric with respect to plane P_S. Since it is a question of scalar fields, the symmetry is expressed by (note that we have set $\xi^1 = x$, $\xi^2 = y$)

$$T_\alpha(x,y) = T_\alpha(-x,y) \ , \ \alpha = 1,2. \tag{6.4.1}$$

Remark 6.4.1 :

In fact, we have assumed in statement (5.4.6) that the fields $T_\alpha \in L^2(\Omega)$, $\alpha = 1,2$. It follows that point values indicated in (6.4.1) may not be necessarily defined. In order to be more rigorous, we need only use classical density arguments ; i.e., first establish the result for sequences of continuous functions such as those in $\mathcal{C}_C(\Omega)$ (the space of continuous functions with compact support in Ω) and then pass to the limit in $L^2(\Omega)$ making use of the fact that $\mathcal{C}_C(\Omega)$ is dense in $L^2(\Omega)$. The same remark can be applied to relations (6.4.2) to (6.4.9) below.

∎

Then, since the data are symmetric with respect to the plane P_S, the same property arises for the displacement field $\vec{u}$ of the middle surface. In order to translate these symmetry conditions in terms of the covariant components u_i of the displacement, i.e.,

$$\vec{u}(x,y) = u_i(x,y)\ \vec{a}^i(x,y) \ , \tag{6.4.2}$$

let us begin to consider the components of the same displacement field on the fixed reference system $(\vec{e_1},\vec{e_2},\vec{e_3})$:

$$\vec{u}(x,y) = \sum_{j=1}^{3} \tilde{u}_j(x,y)\ \vec{e}_j \quad . \tag{6.4.3}$$

The symmetry of the field $\vec{u}$ with respect to plane P_S implies immediately the relations (see Figure 4.2.1)

$$\left.\begin{aligned} \tilde{u}_1(-x,y) &= \tilde{u}_1(x,y) \\ \tilde{u}_2(-x,y) &= -\tilde{u}_2(x,y) \\ \tilde{u}_3(-x,y) &= \tilde{u}_3(x,y) \end{aligned}\right\} \tag{6.4.4}$$

Let us find relations of this type between $u_i(x,y)$ and $u_i(-x,y)$. First, relations (1.1.3), (1.1.6), (6.4.2), and $\vec{a}_3 = \vec{a}^3$ imply

$$\left.\begin{aligned} u_i(x,y) &= \vec{u}(x,y).\vec{a_i}(x,y) \\ u_i(-x,y) &= \vec{u}(-x,y).\vec{a_i}(-x,y) \ . \end{aligned}\right\} \tag{6.4.5}$$

In these relations, we substitute for $\vec{u}$ the relations given in (6.4.3) :

$$\left.\begin{aligned} u_i(x,y) &= \sum_{j=1}^{3} \tilde{u}_j(x,y)\ \vec{a_i}(x,y).\vec{e_j} \\ u_i(-x,y) &= \sum_{j=1}^{3} \tilde{u}_j(-x,y)\ \vec{a_i}(-x,y).\vec{e_j} \quad . \end{aligned}\right\} \tag{6.4.6}$$

For i = 1, relations (4.3.1), (4.3.2) lead to

$$\vec{a_1}(-x,y).\vec{e_1} = -\vec{a_1}(x,y).\vec{e_1} ,$$
$$\vec{a_1}(-x,y).\vec{e_2} = \vec{a_1}(x,y).\vec{e_2} ,$$
$$\vec{a_1}(-x,y).\vec{e_3} = \vec{a_1}(x,y).\vec{e_3} = 0 ,$$

hence, with relations (6.4.4), (6.4.5),

$$u_1(-x,y) = -u_1(x,y). \qquad (6.4.7)$$

Similarly, we can prove that

$$u_2(-x,y) = u_2(x,y) \qquad (6.4.8)$$

$$u_3(-x,y) = u_3(x,y) \qquad (6.4.9)$$

Practically, we use the property of antisymmetry (6.4.7) and properties of symmetry (6.4.8) and (6.4.9), to obtain a new formulation of the problem *equivalent* to the variational formulation (5.4.6) ; but now, the new formulation is set on the *half-domain* Ω_1 – see Figure 6.4.1 – .

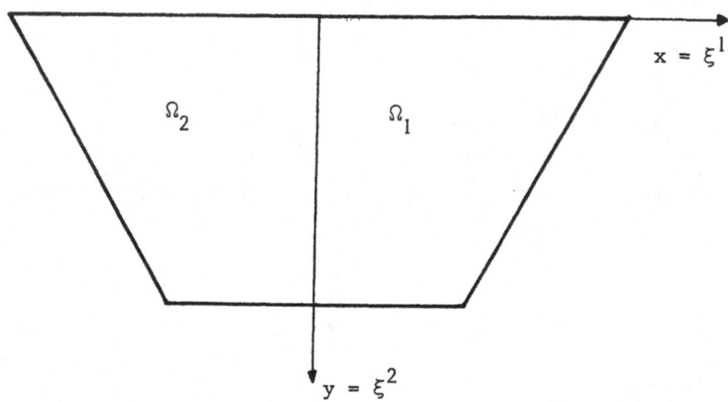

<u>Figure 6.4.1</u> : <u>Partition of domain</u> $\bar{\Omega} = \bar{\Omega}_1 \cup \bar{\Omega}_2$

The following lemma will permit us to express the space $\vec{V}$ as the direct sum of two subspaces having interesting symmetry or antisymmetry properties. Let us set :

$$\vec{\tilde{V}} = \left\{ \vec{v} \in \vec{V},\ \vec{v} = v_i \vec{a}^i,\ v_1 \text{ antisymmetric}/_{x=0}\ ,\ v_2 \text{ and } v_3 \text{ symmetric}/_{x=0} \right\} \tag{6.4.10}$$

$$\vec{\tilde{\tilde{V}}} = \left\{ \vec{v} \in \vec{V},\ \vec{v} = v_i \vec{a}^i,\ v_1 \text{ symmetric}/_{x=0}\ ,\ v_2 \text{ and } v_3 \text{ antisymmetric}/_{x=0} \right\} \tag{6.4.11}$$

<u>Lemma 6.4.1</u> : *The space* $\vec{V}$ *defined in* (5.4.3) *can be written*

$$\vec{V} = \vec{\tilde{V}} \oplus \vec{\tilde{\tilde{V}}}\ , \tag{6.4.12}$$

i.e., for all $\vec{v} \in \vec{V}$, *there exist a unique decomposition*

$$\vec{v} = \vec{\tilde{v}} + \vec{\tilde{\tilde{v}}}\quad ,\quad \vec{\tilde{v}} \in \vec{\tilde{V}}\quad ,\quad \vec{\tilde{\tilde{v}}} \in \vec{\tilde{\tilde{V}}}. \tag{6.4.13}$$

<u>Proof</u> :

This result is immediate if we observe that every component v_i, $i = 1,2,3$, of function $\vec{v}$, can be decomposed in a unique way in its symmetrical part and in its antisymmetrical part. For instance,

$$v_1(x,.) = \frac{1}{2}\,[v_1(x,.) + v_1(-\,x,.)] + \frac{1}{2}\,[v_1(x,.) - v_1(-\,x,.)]\ . \quad \blacksquare$$

The following two lemmas give symmetry properties of matrices $[\mathbf{A}_{IJ}]$ and $\mathbf{F}$ defined in (1.5.10) and (5.5.2), respectively.

<u>Lemma 6.4.2</u> : *The symmetry of middle surface* S *with respect to plane* P_S *involves that matrix* $[\mathbf{A}_{IJ}]$ *is symmetric with respect to axis* x = 0.

<u>Proof</u> :

The following results can be verified using the expressions of geometrical parameters given in section 4.3 :

$$\left.\begin{array}{l} \text{Symmetrical parameters with respect to axis } x = 0 : \\ a_{11},\ a_{22},\ a,\ a^{11},\ a^{22},\ b_{11},\ b_{22},\ b_1^1,\ b_2^2 \\ \Gamma_{12}^1,\ \Gamma_{11}^2,\ \Gamma_{22}^2 \end{array}\right\} \tag{6.4.14}$$

$$\left.\begin{array}{l}\text{Antisymmetrical parameters with respect to axis } x = 0 : \\ a_{12},\ a^{12},\ b^1_2,\ b^2_1,\ b^{12},\ \Gamma^1_{11},\ \Gamma^1_{22},\ \Gamma^2_{12}\end{array}\right\} \quad (6.4.15)$$

$$\left.\begin{array}{l}\text{Zero parameter} \\ b_{12} = 0\end{array}\right\} \quad (6.4.16)$$

From expressions (6.4.14) to (6.4.16), we deduce the following results on Λ^α_β, M^α_β defined in (1.5.6) and (1.5.9), respectively :

$$\Lambda^1_1 \text{ or } \Lambda^2_2 = [\text{AS S AS S AS S S 0 0 0 0 0}] \quad (6.4.17)$$

$$\Lambda^1_2 \text{ or } \Lambda^2_1 = [\text{S AS S AS S AS AS 0 0 0 0 0}] \quad (6.4.18)$$

$$M^1_1 \text{ or } M^2_2 = [\text{AS S AS S AS S S AS S S AS S}] \quad (6.4.19)$$

$$M^1_2 \text{ or } M^2_1 = [\text{S AS S AS S AS AS S AS AS S AS}] \quad (6.4.20)$$

Here S (resp. AS) denote parameters which are symmetrical (resp. antisymmetrical) with respect to x = 0. Mechanical parameters E,e are also assumed to be symmetrical with respect to plane P_S ; then, properties (6.4.17) to (6.4.20) involve the result.

∎

Following steps similar to those used to prove Lemma 6.4.2, we obtain for matrix $\mathbf{F}$:

<u>Lemma 6.4.3</u> : *The symmetries of middle surface* S *and of change of temperature fields* T_α *with respect to plane* P_S *(see relation* (6.4.1)*) imply that matrix* $\mathbf{F}$*, defined in* (5.5.2)*, satisfies the following symmetrical properties*

$${}^t\mathbf{F} = [\text{AS S AS S AS S S AS S S AS S}] . \quad (6.4.21)$$

These notations are those used in relations (6.4.17) *to* (6.4.20).

∎

Then, one can verify the following result :

Theorem 6.4.1 :

The symmetries of the middle surface S *and of change of temperature fields* T_α *with respect to plane* P_S *(see relation* (6.4.1)*) imply the equivalence of variational formulation* (5.4.6) *and following formulation :*

$$\left.\begin{array}{l} \textit{For } T_\alpha \in L^2(\Omega) \textit{ satisfying } (6.4.1), \textit{ find } \vec{u} \in \vec{\tilde{V}} \textit{ such that} \\ \displaystyle\int_{\Omega_1^-} ({}^t\mathbf{U}\ [\mathbf{A}_{IJ}] - {}^t\mathbf{F})\ \mathbf{V}\ dx\ dy = 0 \quad , \forall \vec{\tilde{v}} \in \vec{\tilde{V}} , \end{array}\right\} \quad (6.4.22)$$

where space $\vec{\tilde{V}}$ *is defined in* (6.4.10).

Proof (in two steps) :

$$\text{Step 1 :} \quad \left\{\begin{array}{l} \vec{u} \textit{ solution of } (5.4.6) \\ \\ T_\alpha \quad \textit{symmetric} \end{array}\right\} \Longrightarrow \vec{u} \textit{ is solution of } (6.4.22).$$

Symmetries of middle surface S and the change of temperature fields T_α with respect to plane P_S imply that solution $\vec{u}$ of problem (5.4.6) belongs to space $\vec{\tilde{V}}$.

Now, let $\vec{v}$ be any function of space $\vec{V}$. Lemma 6.4.1 allows us to write in unique way

$$\vec{v} = \vec{\tilde{v}} + \vec{\tilde{\tilde{v}}} \quad , \vec{\tilde{v}} \in \vec{\tilde{V}} \quad , \quad \vec{\tilde{\tilde{v}}} \in \vec{\tilde{\tilde{V}}} \quad . \qquad (6.4.23)$$

By definition of spaces $\vec{\tilde{V}}$ and $\vec{\tilde{\tilde{V}}}$ (see (6.4.10) and (6.4.11)), matrices $\mathbf{U}$, $\tilde{\mathbf{V}}$, $\tilde{\tilde{\mathbf{V}}}$ defined similarly to (1.5.2) and associated respectively with elements $\vec{u} \in \vec{\tilde{V}}$, $\vec{\tilde{v}} \in \vec{\tilde{V}}$ and $\vec{\tilde{\tilde{v}}} \in \vec{\tilde{\tilde{V}}}$, satisfy the following symmetry properties :

$${}^t\mathbf{U} = [\text{AS} \quad \text{S} \quad \text{AS} \quad \text{S} \quad \text{AS} \quad \text{S} \quad \text{S} \quad \text{AS} \quad \text{S} \quad \text{S} \quad \text{AS} \quad \text{S}] \ , \qquad (6.4.24)$$

$${}^t\tilde{\mathbf{V}} = [\text{AS} \quad \text{S} \quad \text{AS} \quad \text{S} \quad \text{AS} \quad \text{S} \quad \text{S} \quad \text{AS} \quad \text{S} \quad \text{S} \quad \text{AS} \quad \text{S}] \ , \qquad (6.4.25)$$

and

$${}^t\tilde{\tilde{\mathbf{V}}} = [\text{S} \quad \text{AS} \quad \text{S} \quad \text{AS} \quad \text{S} \quad \text{AS} \quad \text{AS} \quad \text{S} \quad \text{AS} \quad \text{AS} \quad \text{S} \quad \text{AS}] \ . \qquad (6.4.26)$$

Then, by combining the results of Lemmas 6.4.2 and 6.4.3 with properties (6.4.24) to (6.4.26) leads to the observation that functions ${}^t\mathbf{U}\,[\mathbf{A}_{IJ}]\,\tilde{\mathbf{V}}$, ${}^t\mathbf{F}\tilde{\mathbf{V}}$ (resp. ${}^t\mathbf{U}\,[\mathbf{A}_{IJ}]\,\tilde{\tilde{\mathbf{V}}}$, ${}^t\mathbf{F}\tilde{\tilde{\mathbf{V}}}$) are symmetrical (resp. antisymmetrical) with respect to axis x = 0. Taking into account the decomposition (6.4.23), we obtain, finally,

$$a(\vec{u},\vec{v}) = 2\int_{\Omega_1} {}^t\mathbf{U}\ [\mathbf{A}_{IJ}]\,\tilde{\mathbf{V}}\,dx\,dy\ ,\ \forall\vec{v} \in \tilde{V}\ , \tag{6.4.27}$$

$$f(\vec{v}) = 2\int_{\Omega_1} {}^t\mathbf{F}\tilde{\mathbf{V}}\ dx\,dy,\quad \forall\vec{v} \in \tilde{V}\ , \tag{6.4.28}$$

or also, $\forall\vec{\tilde{\tilde{v}}} \in \tilde{\tilde{V}}$. Then, it suffices to substitute (6.4.27) and (6.4.28) into (5.4.6) to obtain (6.4.22).

<u>Step 2</u> : $\left\{\begin{array}{l}\vec{u}\ \textit{solution of}\ (6.4.22)\\ \\ T_\alpha\ \textit{symmetric}\end{array}\right\} \Longrightarrow \vec{u}\ \textit{solution of}\ (5.4.6)$

According to step 1, we know that the solution of equation (5.4.6) is also a solution of equation (6.4.22), and that proves the existence of solutions for equation (6.4.22). Now let $\vec{u}^* \in \vec{\tilde{V}}$ one solution of equation (6.4.22) ; then, we show that $\vec{u}^*$ is solution of (5.4.6), hence $\vec{u}^* = \vec{u}$ since the solution of equation (5.4.6) is unique.

Denote by $\vec{v}$ any function of $\vec{V}$; Lemma 6.4.1 implies that it is possible to decompose $\vec{v}$ in a unique way as in (6.4.23). Then,

$$2\int_{\Omega_1} {}^t\mathbf{U}^*\ [\mathbf{A}_{IJ}]\,\tilde{\mathbf{V}}\,dx\,dy = \int_{\Omega} {}^t\mathbf{U}^*\ [\mathbf{A}_{IJ}]\,\mathbf{V}\,dx\,dy = a(\vec{u}^*,\vec{v}),\ \forall\vec{v} \in \vec{V},$$

and

$$2\int_{\Omega_1} {}^t\mathbf{F}\tilde{\mathbf{V}}\ dx\,dy = \int_{\Omega} {}^t\mathbf{F}\mathbf{V}\ dx\,dy = f(\vec{v}),\ \forall\vec{v} \in \vec{V}.$$

By substituting into (6.4.22), we obtain

$$a(\vec{u}^*,\vec{v}) = f(\vec{v}),\ \forall\vec{v} \in \vec{V},$$

that is exactly formulation (5.4.6). ∎

For all practical purposes the symmetry conditions stated in the definition of space $\vec{\vec{V}}$ (see(6.4.10)) introduce *"pseudo-boundary conditions"* on the border x = 0 of domain Ω_1, i.e., for $\vec{u} \in \vec{\vec{V}}$, $\vec{v} \in \vec{\vec{V}}$:

$$ {}^t\mathbf{U}\,(o,y) = [0\ \ 1\ \ 0\ \ 1\ \ 0\ \ 1\ \ 1\ \ 0\ \ 1\ \ 1\ \ 0\ \ 1] \,, \qquad (6.4.29) $$

$$ {}^t\mathbf{V}\,(o,y) = [0\ \ 1\ \ 0\ \ 1\ \ 0\ \ 1\ \ 1\ \ 0\ \ 1\ \ 1\ \ 0\ \ 1] \,, \qquad (6.4.30) $$

where 0 (resp. 1) indicates that corresponding element is zero (resp. free).

In Figures 6.4.2 to 6.4.4, we indicate how to take into account

(i) *"pseudo-boundary conditions"* on border x = 0 of half-domain Ω_1,

(ii) *clamped conditions* on border y = 1 of half-domain Ω_1,

for every finite element space considered in sections 3.2 to 3.6. We note that these spaces are constructed with the help of finite elements defined in Figure 2.1.1 to 2.1.5. In particular, the corner (x = 0, y = 1) combines both types of conditions.

Results which appear in Figures 6.4.2 to 6.4.4 are relative to *global* degrees of freedom and they have to be read as follows :

[0] corresponding degree of freedom is zero ;

[1] corresponding degree of freedom is really an unknown of the problem.

6.5. Solution method

After taking into account boundary conditions and symmetry conditions, we arrive at an equation of the type (3.2.1), i.e., a system of linear equations. The corresponding matrix $\mathbf{A}$ is symmetric and definite positive.

To solve this system, we use a *standard CHOLESKI's method* : Matrix $\mathbf{A}$ is decomposed as the product

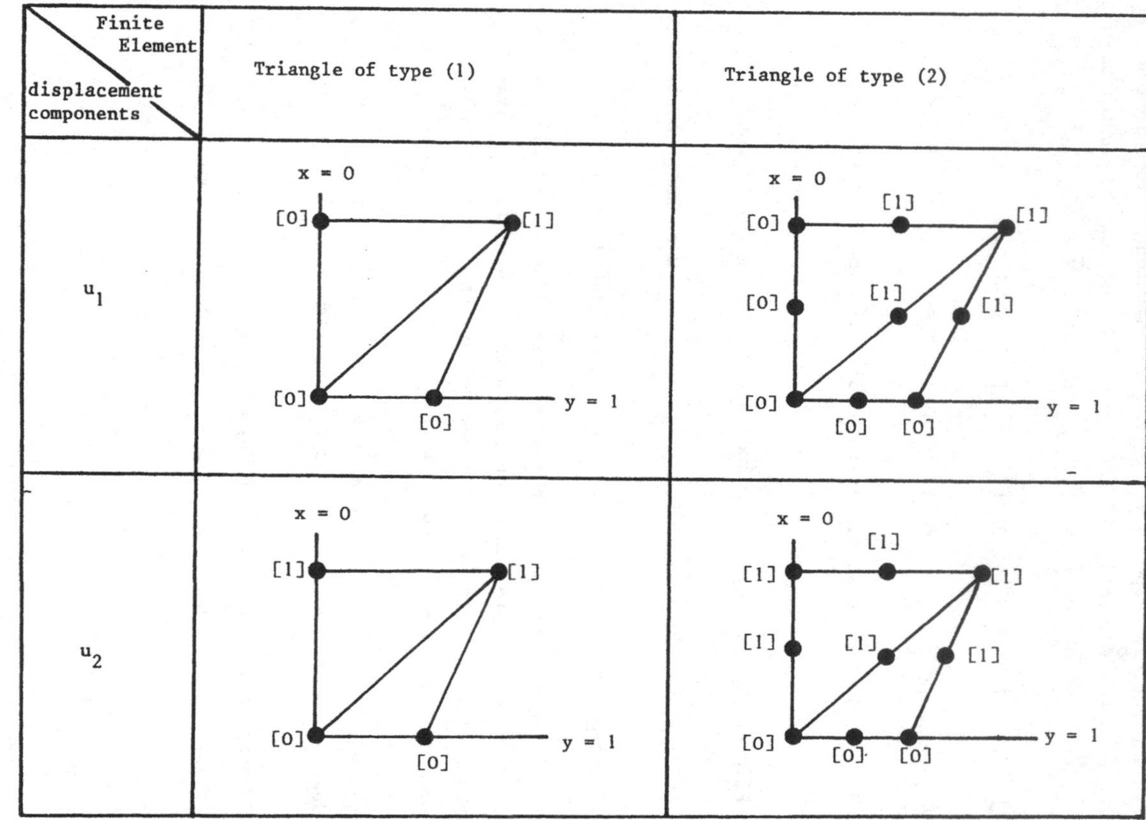

Figure 6.4.2 : Taking into account symmetry conditions (x = 0) and clamped conditions (y = 1) for triangles of type (1) or (2).

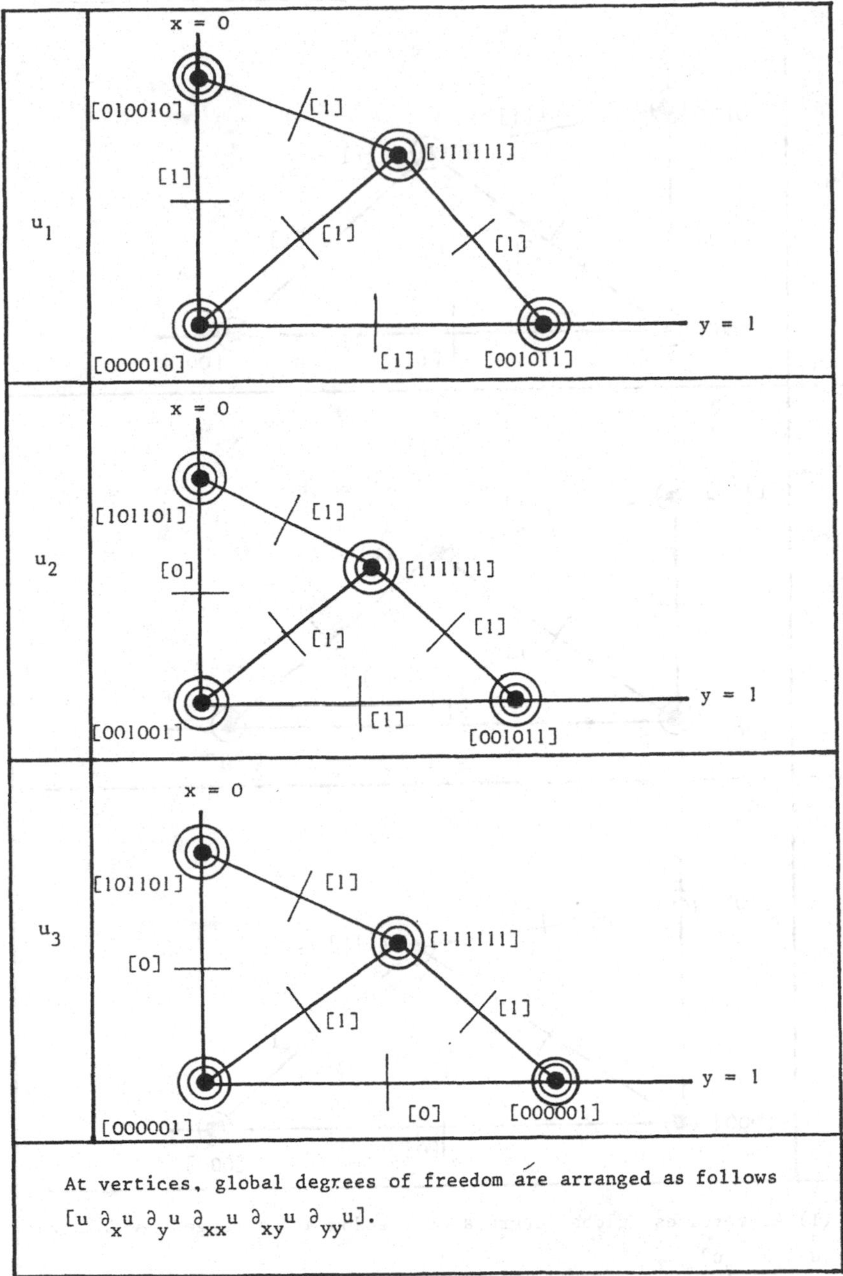

Figure 6.4.3 : Taking into account symmetry conditions ($x = 0$) and clamped conditions ($y = 1$) for the ARGYRIS triangle.

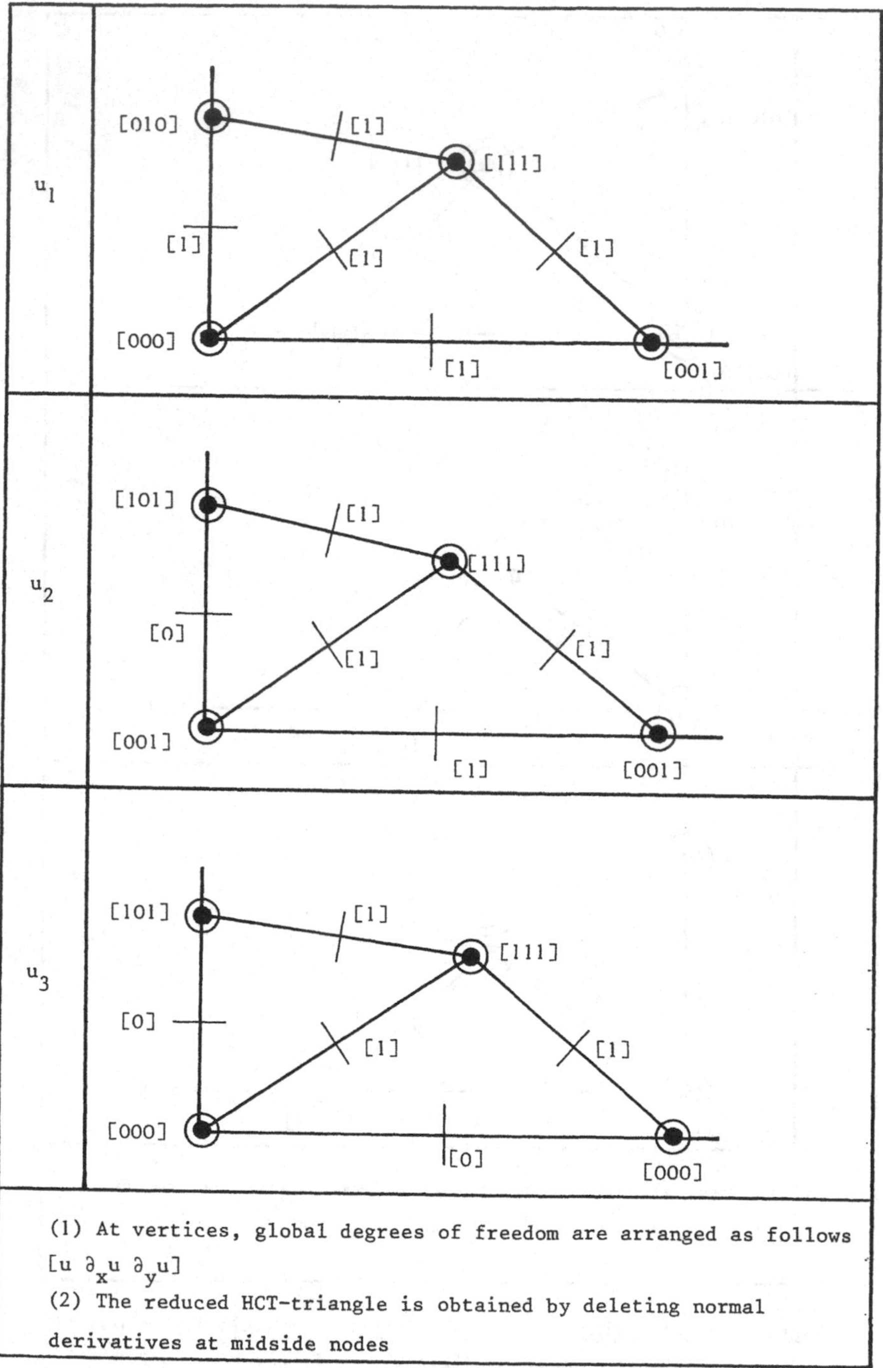

Figure 6.4.4 : Taking into account symmetry conditions (x = 0) and clamped conditions (y = 1) for the complete or reduced HCT-triangles.

$$\mathbf{A} = {}^tSS ,$$

where S denotes an upper-triangular matrix. This method allows us to consider several different second members in the course of only one inversion, so that we are able to systematically consider the effect of different types of loading.

In the present case, very few coefficients of the matrix $\mathbf{A}$ are different from zero. Consequently, we shall use a *sky-line bandwith factorization* in order to reject most of zero terms. Let us define the notion of a *profile* for an upper triangular matrix S (resp. of a square symmetric matrix $\mathbf{A}$) :

Definition 6.5.1 : Let an upper triangular matrix S (resp. a square symmetric matrix $\mathbf{A}$) be given with elements that are denoted S_{ij} (resp. $\mathbf{A}_{ij}$), i,j = 1,...,N. The *profile of matrix* S (resp. $\mathbf{A}$) is the N-line vector P with elements P(j) , j = 1,...,N such that

$$\left.\begin{array}{l} P(j) \in [1,\ldots,j] \; ; \\ \text{if } P(j) \neq 1, \text{ then } \forall i \in [1, P(j) - 1],\ S_{ij} = 0 \text{ (resp. } \mathbf{A}_{ij} = 0) \end{array}\right\} .$$

■

For instance, the profile of the following upper triangular matrix

$$S = \begin{bmatrix} 1 & 0 & 3 & 0 & 0 \\ & 2 & 5 & 0 & 0 \\ & & 0 & 0 & 2 \\ & & & 1 & 0 \\ \mathbf{O} & & & & 4 \end{bmatrix} ,$$

is the 5-line vector

$$P = [1 \quad 2 \quad 1 \quad 4 \quad 3] .$$

Then, the main result that we use is stated in JENNINGS [1971] :

Theorem 6.5.1 :

Let $\mathbf{A}$ *be a symmetric and positive definite matrix. CHOLESKI's method associates with* $\mathbf{A}$ *an upper triangular matrix* S *such that* $\mathbf{A} = {}^t\mathrm{SS}$. *Then, matrices* $\mathbf{A}$ *and* S *have the same profile.*

■

6.6. Calculation of the displacements

The solution method described in section 6.5 gives the values of the unknown degrees of freedom. Then, using boundary conditions and symmetry conditions, we know the values of all global or border degrees of freedom for the finite element method considered.

Next, consider any point of the dam of coordinates ($\xi^1 = x$, $\xi^2 = y$, ξ^3). First, the point (ξ^1,ξ^2) of $\bar{\Omega}$ belongs to (at least) one triangle $K \in \mathcal{T}_h$. In this triangle, point (ξ^1,ξ^2) has $(\lambda_1,\lambda_2,\lambda_3)$ as barycentric coordinates. According to the finite element method used, relations (3.2.5), (3.3.1), (3.4.1), (3.5.1), or (3.6.1), allow us to define approximate values of displacement components as well as values of some of their partial derivatives *at point* $\vec{\phi}(\xi^1,\xi^2)$ *of the middle surface of the dam*, i.e.,

$$\left.\begin{aligned} {}^t\mathbf{U}_h = {}^t[&u_{1h}\ u_{1h,1}\ u_{1h,2}\ u_{2h}\ u_{2h,1}\ u_{2h,2} \\ &u_{3h}\ u_{3h,1}\ u_{3h,2}\ u_{3h,11}\ u_{3h,12}\ u_{3h,22}]\ . \end{aligned}\right\} \tag{6.6.1}$$

Next, relation (1.3.20) allow us to find an approximation of the *displacement* of the point (ξ^1,ξ^2,ξ^3) of the dam, i.e.,

$$\vec{U}_h = \left\{u_{\alpha h} - \xi^3(u_{3h,\alpha} + b^\lambda_\alpha\, u_{\lambda h})\right\}\ \vec{a}^\alpha + u_{3h}\ \vec{a}^3\ . \tag{6.6.2}$$

6.7. Calculation of the stresses at any point of the dam ; physical components

The calculation of approximate values of the components of the stress tensor at any point (ξ^1,ξ^2,ξ^3) of the dam involves the three following steps :

Step 1 : Calculation of the displacement at point ($\xi^1 = x$, $\xi^2 = y$)

According to (6.6.1), we compute matrix ${}^t\mathbf{U}_h$ at point (ξ^1,ξ^2).

<u>Step 2</u> : <u>Calculation of components</u> γ^{*i}_{j} <u>of strain tensor at point (ξ^1,ξ^2,ξ^3) of the dam</u> :

From matrix ${}^t\mathbf{U}_h$, relations (1.5.5) and (1.5.8) make it possible to deduce values of components γ^{α}_{β} and $\bar{\rho}^{\alpha}_{\beta}$ of the strain and change of curvature tensors of the middle surface at point $\vec{\phi}(\xi^1,\xi^2)$ of the middle surface, i.e.,

$$\gamma^{\alpha}_{\beta}(\vec{u_h}) = \Lambda^{\alpha}_{\beta}\mathbf{U}_h \,, \tag{6.7.1}$$

$$\bar{\rho}^{\alpha}_{\beta}(\vec{u_h}) = M^{\alpha}_{\beta}\mathbf{U}_h \,. \tag{6.7.2}$$

Then, by using relations (1.3.8) to (1.3.11) and (5.3.15), we obtain an approximation of components γ^{*i}_{j} of *strain tensor* at point (ξ^1,ξ^2,ξ^3) of the dam, i.e., just keeping the linear part in ξ^3 :

$$\left.\begin{aligned}
\gamma^{*\beta}_{\alpha} &= g^{\beta i}\,\gamma^{*}_{i\alpha} = \gamma^{\beta}_{\alpha} - \xi^3\,[\bar{\rho}^{\beta}_{\alpha} - 2\,b^{\beta}_{\lambda}\,\gamma^{\lambda}_{\alpha}] \\
\gamma^{*3}_{\alpha} &= g^{3i}\,\gamma^{*}_{i\alpha} = \gamma^{*}_{3\alpha} = 0 \\
\gamma^{*\beta}_{3} &= g^{\beta i}\,\gamma^{*}_{3i} = 0 \\
\gamma^{*3}_{3} &= -\frac{\nu}{1-\nu}\,\gamma^{*\alpha}_{\alpha}
\end{aligned}\right\} \tag{6.7.3}$$

<u>Step 3</u> : <u>Calculation of components</u> σ^{*ij} <u>of stress tensor at point (ξ^1,ξ^2,ξ^3) of the dam</u> :

Relations (5.3.1), (5.3.2), (5.3.10), and (5.3.12) imply

$$\left.\begin{aligned}
\sigma^{*ij} = {}& \frac{E}{2(1+\nu)}\,[g^{ik}\,\gamma^{*j}_{k} + g^{jk}\,\gamma^{*i}_{k} + \frac{2\nu}{1-2\nu}\,g^{ij}\,\gamma^{*k}_{k}] \\
& - \frac{E\bar{\alpha}}{1-2\nu}\,(T_1 + \xi^3 T_2)\,g^{ij} \,.
\end{aligned}\right\} \tag{6.7.4}$$

Combining the above relation with relations (6.7.1) to (6.7.3), we completely determine the contravariant components σ^{*ij}.

∎

Definition of physical components :

Generally, covariant or contravariant components of a vector or of a tensor, referred to a general curvilinear coordinate system, *do not have natural physical dimensions of the field in consideration*. This is the reason we introduce the notion of *physical components* of a vector or of a tensor. We consider two kind of situations :

(i) *Curvilinear coordinate system is orthogonal, i.e.,* vectors of local basis $\vec{g_i}$ are orthogonal

$$\vec{g_i}\cdot\vec{g_j} = 0 \quad \text{if } i \neq j\ . \qquad (6.7.5)$$

In this case, it is costumary to use the following definition of physical components : these are components - not tensorial - of the tensor considered in the orthonormal basis $(\frac{\vec{g_i}}{\|\vec{g_i}\|})$, i = 1,2,3. For more details, we refer for example to ERICKSEN [1960, chapter II, pages 797-805], NAGHDI [1972, appendix A4, pages 631-633], TRUESDELL [1953] whose results are in agreement with those of Mc CONNELL [1931] who was precursory in this matter.

(ii) *Curvilinear coordinate system is general :*

In this case, TRUESDELL [1953, pages 346-347] mentions the absence of unity in the definitions used by different authors. In the following, we use the definitions of TRUESDELL [1953] which are also proposed by MALVERN [1969, pp. 606-613]. Then, to obtain physical components of the stress tensor at any point of the dam from corresponding contravariant components σ^{*ij}, we proceed as follows :

Step 1 : Definition of mixed components :

Since tensor (σ^{*kl}) is symmetric, mixed components are defined by $(\sigma^{*i.}_{\ \ .j} = \sigma^{*.i}_{\ \ j.} = \sigma^{*i}_{\ \ j})$

$$\sigma^{*i}_{\ \ j} = g_{jk}\,\sigma^{*ik}\ , \qquad (6.7.6)$$

or, by combining with relations (6.7.4),

$$\sigma^{*i}_{\ j} = \frac{E}{1+\nu} [\gamma^{*i}_{\ j} + \frac{\nu}{1-2\nu} \delta^i_j \gamma^{*\ell}_{\ \ell}] - \frac{E\bar{\alpha}}{1-2\nu} (T_1 + \xi^3 T_2) \delta^i_j \qquad (6.7.7)$$

Finally, with relations (6.7.3), we obtain :

$$\left.\begin{aligned} \sigma^{*\beta}_{\ \omega} &= \frac{E}{1+\nu} [\gamma^{\beta}_{\omega} + \frac{\nu}{1-\nu} \delta^{\beta}_{\omega} \gamma^{\varepsilon}_{\varepsilon} \\ &\quad - \xi^3 \left\{ \bar{\rho}^{\beta}_{\omega} - 2b^{\beta}_{\lambda} \gamma^{\lambda}_{\omega} + \frac{\nu}{1-\nu} \delta^{\beta}_{\omega} (\bar{\rho}^{\varepsilon}_{\varepsilon} - 2b^{\varepsilon}_{\lambda} \gamma^{\lambda}_{\varepsilon}) \right\}] \\ &\quad - \frac{E\bar{\alpha}}{1-2\nu} (T_1 + \xi^3 T_2) \delta^{\beta}_{\omega} \\ \sigma^{*3}_{\ \omega} &= \sigma^{*\beta}_{\ 3} = 0 \\ \sigma^{*3}_{\ 3} &= - \frac{E\bar{\alpha}}{1-2\nu} (T_1 + \xi^3 T_2) \end{aligned}\right\} \qquad (6.7.8)$$

<u>Step 2</u> : <u>Definition of (right-) physical components</u> :

According to TRUESDELL [1953, (5.10)], (right-) physical components of stress tensor are given by

$$\sigma^*(ij) = \sqrt{\frac{g_{ii}}{g_{jj}}}\ \sigma^{*i}_{\ j} \ , \ (i,j \text{ unsummed}) \qquad (6.7.9)$$

From (1.3.7), we get

$$\left.\begin{aligned} \sqrt{g_{\beta\beta}} &= \sqrt{a_{\beta\beta}} \left(1 - \frac{b_{\beta\beta}}{a_{\beta\beta}} \xi^3\right) , \ (\beta \text{ unsummed}) \\ \sqrt{g_{33}} &= 1 , \end{aligned}\right\} \qquad (6.7.10)$$

retaining only the linear part in ξ^3.

Finally, by combining (6.7.8) to (6.7.10), we obtain :

$$\left.\begin{aligned}
\sigma^*(11) &= \frac{E}{1-\nu^2}\left[\gamma^1_1 + \nu\gamma^2_2 - \xi^3\left\{\bar{\rho}^1_1 + \nu\bar{\rho}^2_2 - 2(b^1_\lambda\,\gamma^\lambda_1 + \nu b^2_\lambda\,\gamma^\lambda_2)\right\}\right] \\
&\quad - \frac{E\bar{\alpha}}{1-2\nu}(T_1 + \xi^3 T_2) \\
\sigma^*(22) &= \frac{E}{1-\nu^2}\left[\nu\gamma^1_1 + \gamma^2_2 - \xi^3\left\{\nu\bar{\rho}^1_1 + \bar{\rho}^2_2 - 2(\nu b^1_\lambda\,\gamma^\lambda_1 + b^2_\lambda\,\gamma^\lambda_2)\right\}\right] \\
&\quad - \frac{E\bar{\alpha}}{1-2\nu}(T_1 + \xi^3 T_2) \\
\sigma^*(33) &= -\frac{E\bar{\alpha}}{1-2\nu}(T_1 + \xi^3 T_2) \\
\sigma^*(12) &= \frac{E}{1+\nu}\sqrt{\frac{a_{11}}{a_{22}}}\left[\gamma^1_2 + \xi^3\left(\frac{b_{22}}{a_{22}} - \frac{b_{11}}{a_{11}}\right)\gamma^1_2 - \xi^3\,(\bar{\rho}^1_2 - 2b^1_\lambda\,\gamma^\lambda_2)\right] \\
\sigma^*(21) &= \frac{E}{1+\nu}\sqrt{\frac{a_{22}}{a_{11}}}\left[\gamma^2_1 - \xi^3\left(\frac{b_{22}}{a_{22}} - \frac{b_{11}}{a_{11}}\right)\gamma^2_1 - \xi^3\,(\bar{\rho}^2_1 - 2b^2_\lambda\,\gamma^\lambda_1)\right] \\
\sigma^*(13) &= \sigma^*(23) = \sigma^*(31) = \sigma^*(32) = 0
\end{aligned}\right\} \quad (6.7.11)$$

<u>Remark 6.7.1</u> :

Symmetry of matrix σ^{*ij} does not involve symmetry of matrix $\sigma^*(ij)$.

■

<u>Remark 6.7.2</u> :

According to the terminology of TRUESDELL [1953, § 5], we have defined, in (6.7.9), (right-) physical components. In the same way, one can define (left-) physical components (TRUESDELL [1953, (5.10)]) :

$$(ji)\sigma^* = \sqrt{\frac{g_{ii}}{g_{jj}}}\,\sigma^{*i}_{\ j} \quad , \text{ (i,j unsummed)} \qquad (6.7.12)$$

Thus, since tensor σ^{*ij} is symmetric, matrices $[(ij)\sigma^*]$ and $[\sigma^*(ij)]$ are transposed one from the other.

■

6.8. Calculation of the stresses on the upstream and downstream walls of the dam ; physical components

By definition, upstream and downstream walls of the dam are respectively obtained for $\xi^3 = \frac{e}{2}$ and $\xi^3 = -\frac{e}{2}$. Then, by substituting these values of ξ^3 into relations (6.7.8) and (6.7.11), we obtain the values of mixed components and (right-) physical components of stress tensor along the external walls of the dam.

CHAPTER 7

NUMERICAL EXPERIMENTS

Orientation :

In order to illustrate the previous considerations, we give some numerical experiments derived using the ARGYRIS triangle. In section 7.1, we examine separate effects of some specific loads such as *the weight of the dam, water pressure,* and *change of temperature*. Next, we consider the corresponding effect of the combination of these loads. All these results are obtained for a regular triangulation involving 32 elements. The effects of *changes of triangulation* on the behaviour of the computed solution are examined in section 7.2. Finally, in section 7.3, we consider the effect of *changes of numerical integration scheme*.

These numerical experiments are very interesting because

(i) they can be *qualitatively compared* with the experimental results given in RYDZEWSKI [1965, p. 639] for a similar arch dam ;

(ii) they are *very stable* with respect to perturbations such as changes of triangulations or changes of numerical integration scheme.

7.1. The effect of different kinds of loads

According to specifications of section 6.4, we work on the *half* domain Ω. With notations of Figure 6.2.1, we consider the triangulation indicated in Figure 7.1.1 which corresponds to $M = N = 4$.

Then, we consider successively different kinds of load effects. For each case Fig. 7.1.2 to 7.1.9 show the distribution of stresses on the upstream and downstream faces of the arch dam. We note that the level of water in the reservoir is assumed to be $Z_1 = 5$ meters (see (5.2.1) ; we record that the origin is at the top of the dam) meanwhile $E = 2.10^4$ M.New/m^2, $\nu = 0.2$.

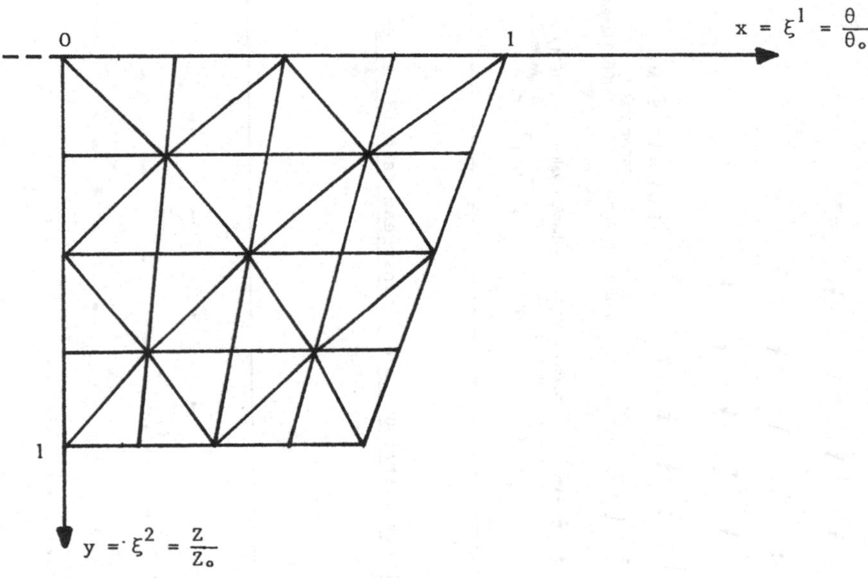

Figure 7.1.1 : Triangulation of the half domain Ω (M = N = 4)

Effect of gravitational loading :

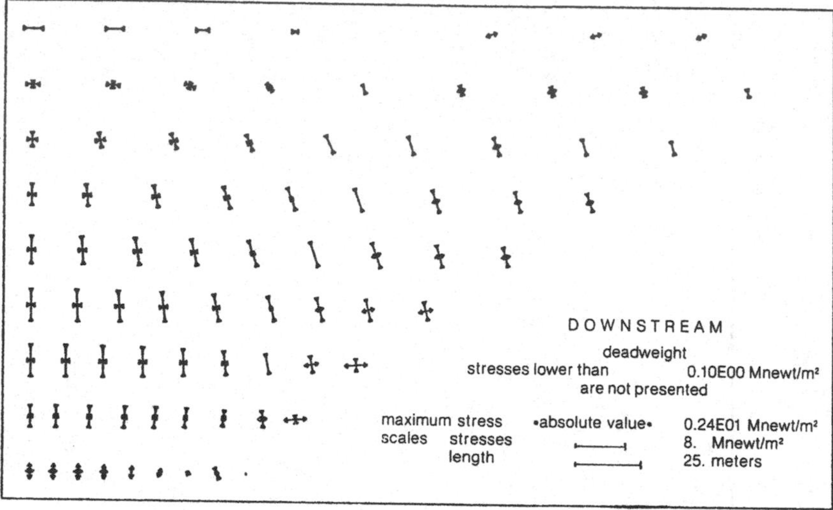

Figure 7.1.2 : Stress distribution on downstream face (32 triangles)

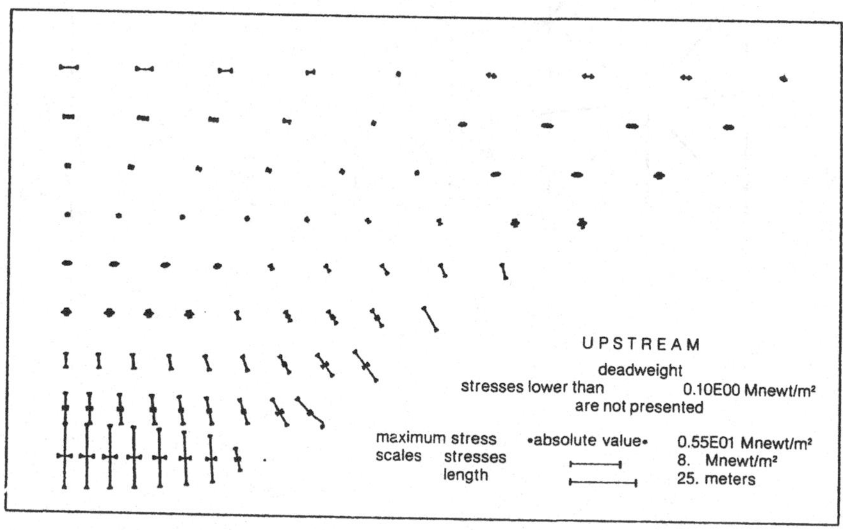

Figure 7.1.3 : Stress distribution on upstream face (32 triangles)

Effect of hydrostatic loading :

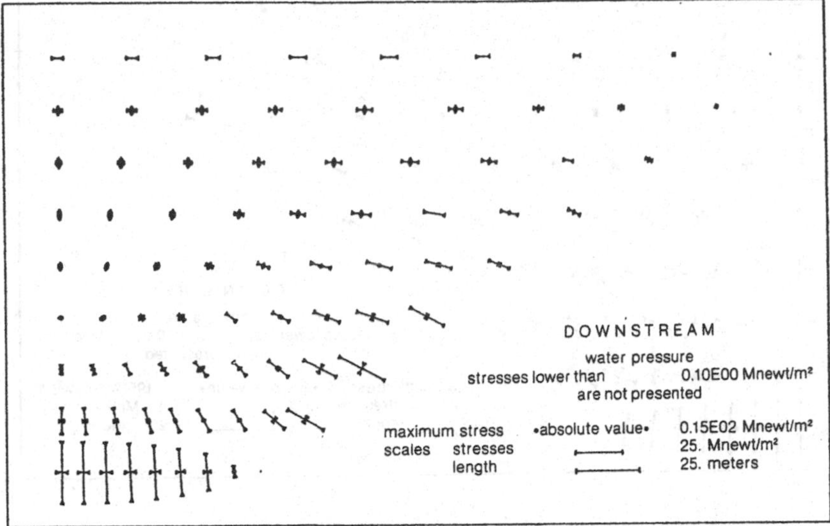

Figure 7.1.4 : Stress distribution on downstream face (32 triangles)

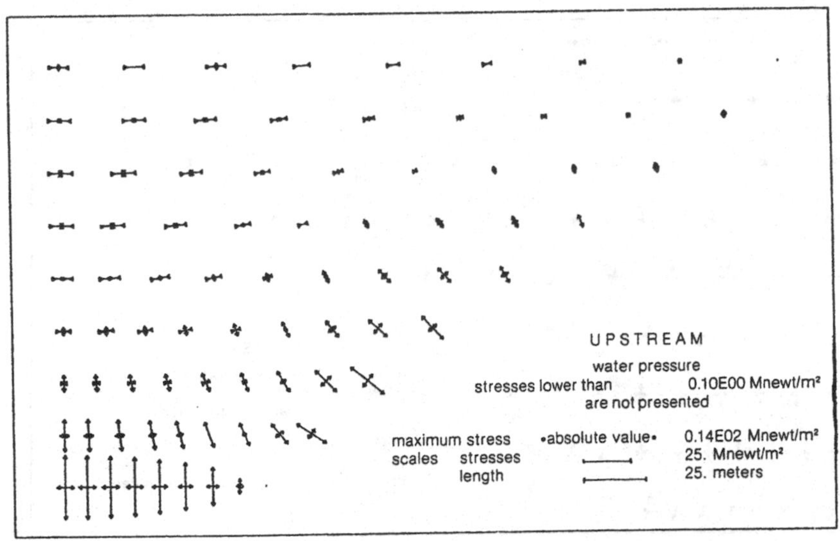

Figure 7.1.5 : Stress distribution on upstream face (32 triangles)

Effect of changes of temperature

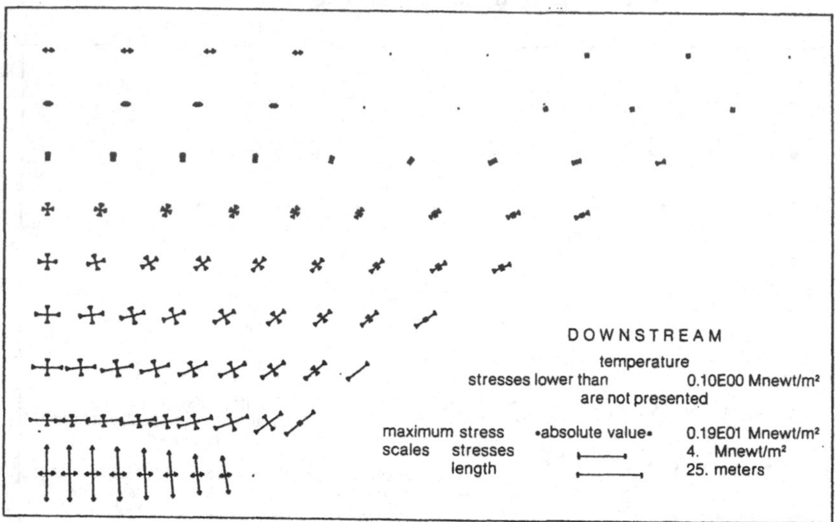

Figure 7.1.6 : Stress distribution on downstream face (32 triangles)

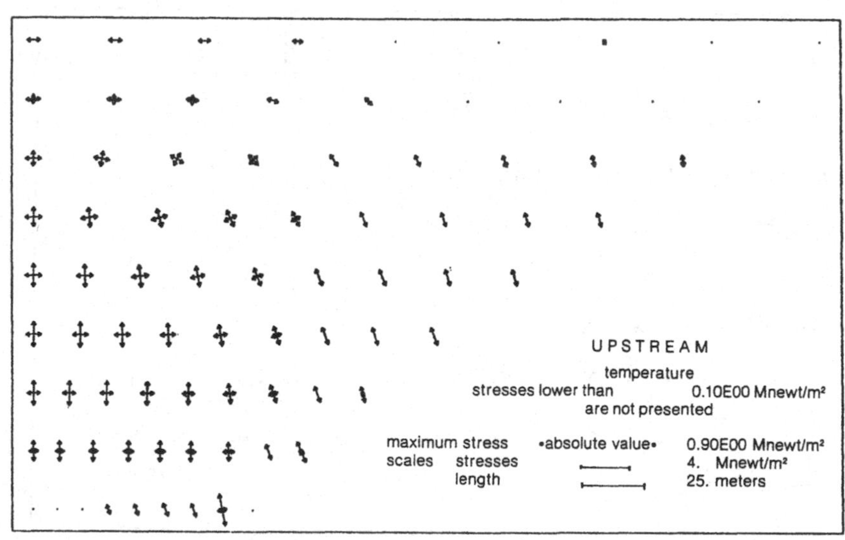

Figure 7.1.7 : Stress distribution on upstream face (32 triangles)

Combined effect of previous loads :

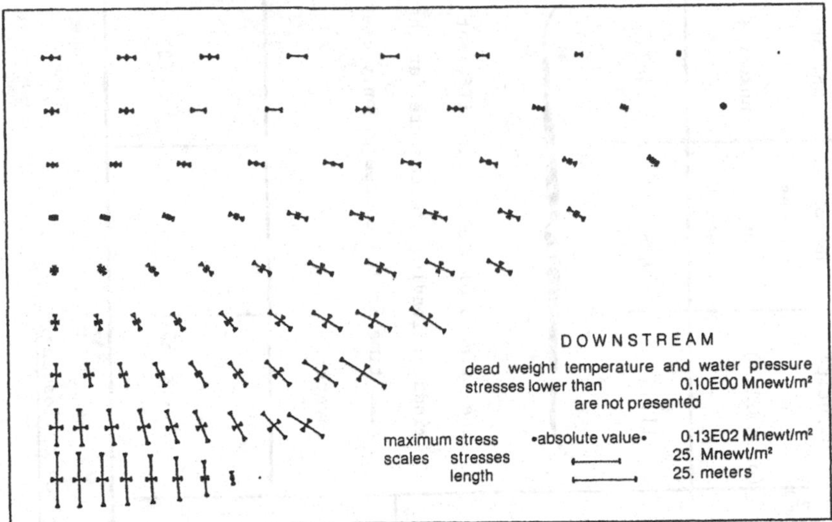

Figure 7.1.8 : Stress distribution on downstream face (32 triangles)

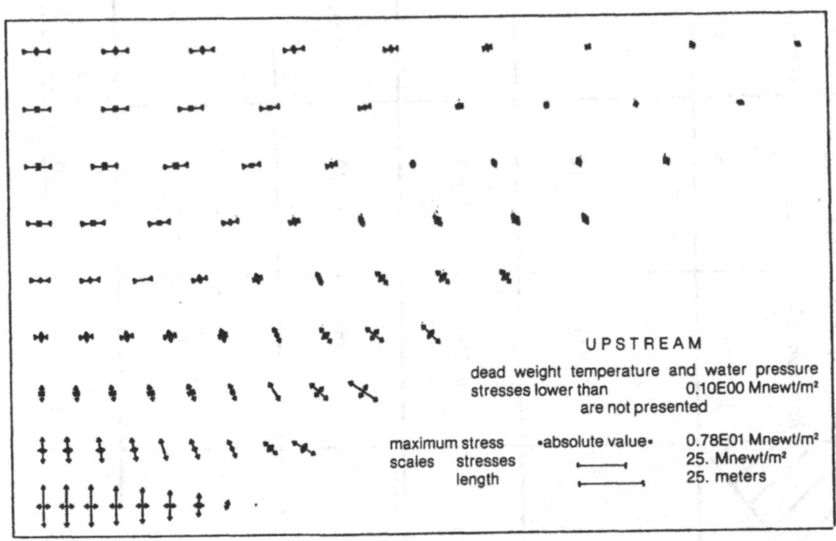

Figure 7.1.9 : Stress distribution on upstream face (32 triangles)

Loads	Top of crown cantilever- *Horizontal stresses* (MN/m^2)		Base of crown cantilever- *Horizontal stresses* (MN/m^2)		Computing time (on CRAY 1)	Nb. of degrees of freedom	Nb. of unknowns
	Upstream	Downstream	Upstream	Downstream			
Gravitational	- 1.374	- 1.512	- 1.107	0.142	11.45	618	463
Hydrostatic	- 4.800	- 2.883	2.753	- 2.974	The effects of these four different kinds of loading are computed at the same time in the course of only one inversion.		
Thermal	0.258	0.179	0.015	0.382			
Combined effects	- 5.916	- 4.216	1.662	- 2.450	11.4 s	618	463

Table 7.1.1 : Effect of different kinds of loading

With notations of section 5.3, the effect of changes of temperature is associated to the following data :

$$T_{up}(\xi^1,\xi^2) = 0 \ , \ T_{do}(\xi^1,\xi^2) = \begin{cases} 0 \text{ if } 0 \le \xi^2 \le \dfrac{Z_1}{Z_o} \\ \dfrac{Z_o\xi^2 - Z_1}{Z_o - Z_1} 16^{\circ} \text{ if } \dfrac{Z_1}{Z_o} \le \xi^2 \le 1 \end{cases} \tag{7.1.1}$$

Obviously, we can consider other kinds of thermal loading. This choice is done for simplicity : according to Remark 5.3.3, we do not have to add a boundary term at the top of the dam.

In table 7.1.1, we summarize the values of *horizontal* components of physical stresses at top and base of crown cantilever for upstream and downstream faces. We also mention the computing time, the number of degrees of freedom and the number of unknowns.

From now on, we shall consider results of Figures 7.1.8 and 7.1.9 as *reference results*. In two next sections, we shall compare these results with those obtained from changes of triangulation or changes of numerical integration scheme.

7.2. The effect of changes of triangulation

The results of section 7.1 have been obtained from the triangulation described in Figure 7.1.1. The Figures 7.2.2 to 7.2.9 present the results obtained from triangulations corresponding to M = N = 2,3,5 and 8. Moreover, in Figures 7.2.10 and 7.2.11, we give the results obtained from the "irregular" mesh of Figure 7.2.1.

These meshes are obtained from a given value of parameter $\varepsilon (0 < \varepsilon < 0.25)$ by dividing the interval [0,1] in M segments $(M \ge 1)$ whose lengths are successively $a, a^{1+\varepsilon}, \ldots, a^{1+(M-1)\varepsilon}$. In other words, a is solution of the equation

$$a = \frac{1 - a^{\varepsilon}}{1 - a^{M\varepsilon}} \quad .$$

Figure 7.2.1 corresponds to the case M = 4 for both intervals $x,y \in [0,1]$ and $\varepsilon = 0.15$. The interest of this irregular mesh lies in the refinement closed to the clamped part of the dam where large stresses appear.

In Table 7.2.1, we summarize the values of *horizontal* components of physical stresses at top and base of crown cantilever for upstream and downstream faces. We also mention the computing time, the number of degrees of freedom and the number of unknowns.

7.3. The effect of changes of numerical integration scheme

In sections 2.3 and 2.4, we have noted that for an approximation using the ARGYRIS triangle, we have to use a scheme exact for polynomials of degree 8. Here, we compare the results obtained from such a scheme (16 points) with the results obtained from schemes exact for polynomials of degree 6 (12 points) on the one hand, and degree 10 (25 points) on the other. These schemes are given by LYNESS and JESPERSEN [1975] for 12 and 16 points and by LAURSEN and GELLERT [1978, formula 15b] for the 25 points.

For both schemes, we obtain results very close to those given in Figures 7.1.8 and 7.1.9 and we merely record these in the Table 7.3.1. In the case of the 12 points scheme, one can attribute this result to the fact that the arch dam is not very thin so that the pure flexural term should be preponderant. In particular, the integration of the term containing $u_{3|\alpha\beta}\, u_{3|\lambda\mu}$ requires only a scheme exact for polynomials of degree 6.

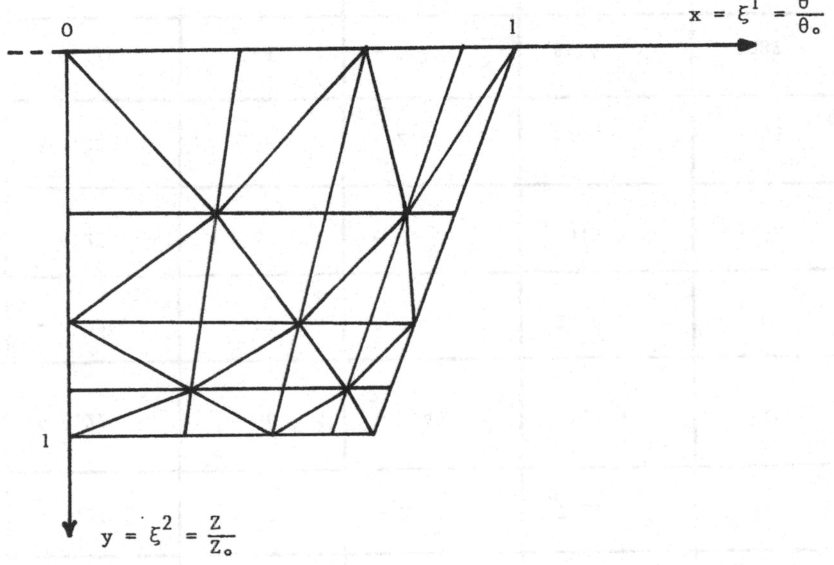

<u>Figure 7.2.1</u> : <u>Irregular mesh</u> ($\varepsilon = 0.15$)

Type of triangulation	Top of crown cantilever- *Horizontal Stresses* (MN/m^2)		Base of crown cantilever- *Horizontal Stresses* (MN/m^2)		Computing time (on CRAY 1)	Nb. of degrees of freedom	Nb. of unknowns
	Upstream	Downstream	Upstream	Downstream			
M = N = 2	- 5.77	- 4.27	1.57	- 2.38	2.1 s	210	123
M = N = 3	- 5.90	- 4.21	1.65	- 2.44	7.3. s	387	266
M = N = 4 (reference)	- 5.91	- 4.22	1.66	- 2.45	11.2 s	618	463
M = N = 5	- 5.92	- 4.22	1.67	- 2.47	22.6 s	903	714
M = N = 6	- 5.93	- 4.22	1.67	- 2.47	34.9 s	1242	1019
M = N = 7	- 5.94	- 4.22	1.68	- 2.47	57.4 s	1635	1378
M = N = 8	- 5.94	- 4.22	1.68	- 2.47	84.4 s	2082	1791
Irregular mesh M = N = 4	- 5.91	- 4.22	1.66	- 2.45	11.2 s	618	463

Table 7.2.1.: Effects of changes of triangulation

Triangulation corresponding to M = N = 2.

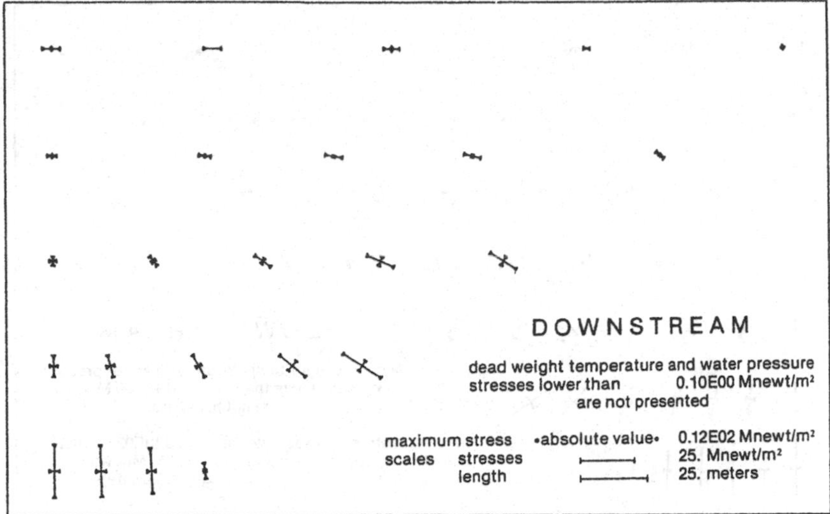

Figure 7.2.2 : Stress distribution on downstream face (8 triangles)
(compare with Figure 7.1.8)

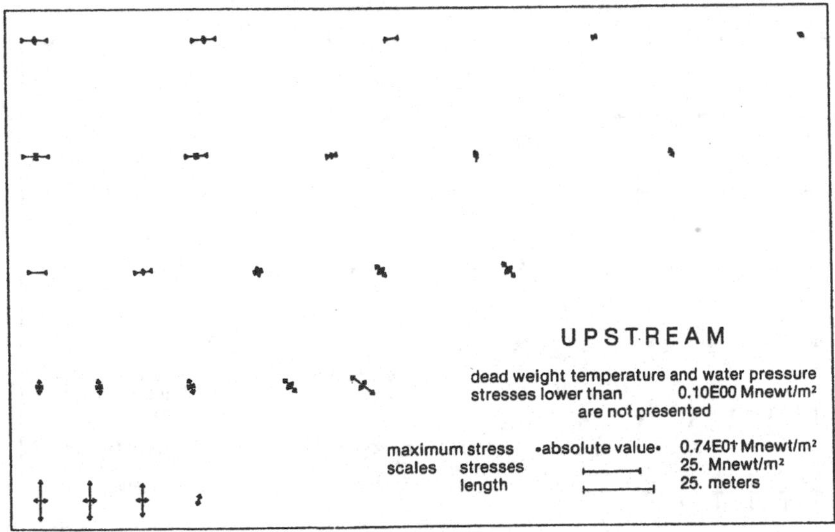

Figure 7.2.3 : Stress distribution on upstream face (8 triangles)
(compare with Figure 7.1.9)

Triangulation corresponding to M = N = 3

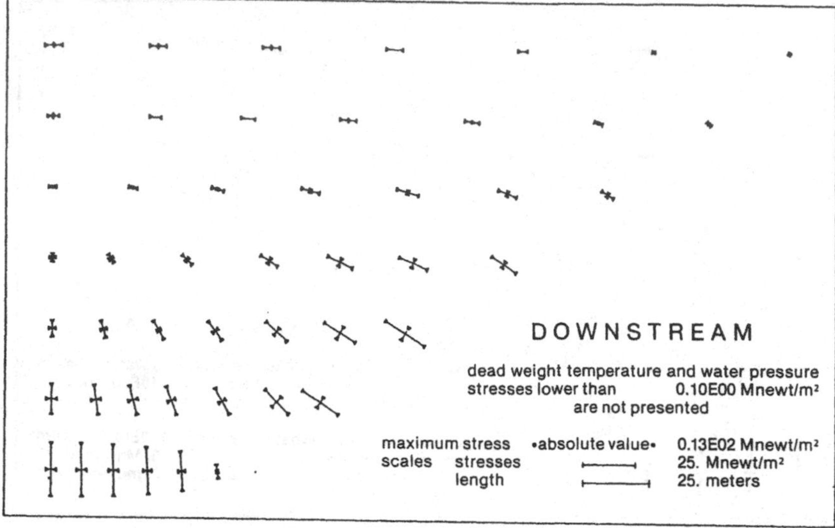

Figure 7.2.4 : Stress distribution on downstream face (18 triangles)
(compare with Figure 7.1.8)

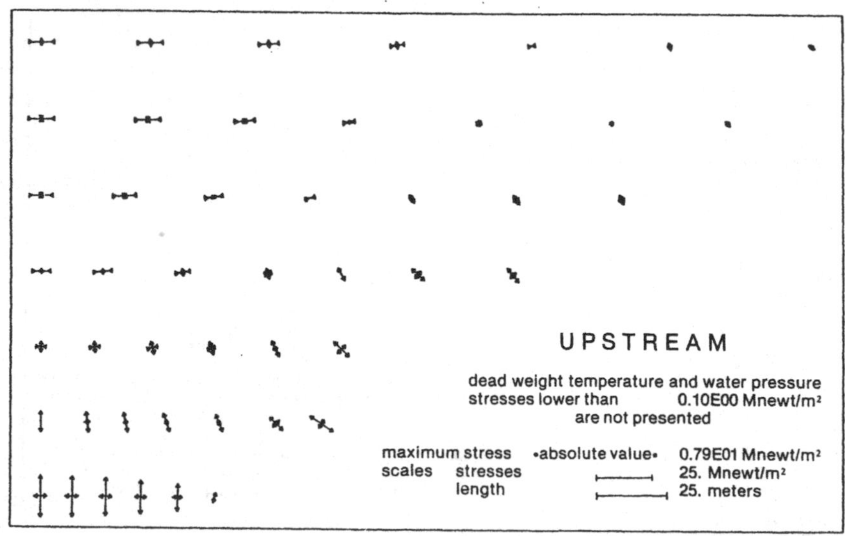

Figure 7.2.5 : Stress distribution on upstream face (18 triangles)
(compare with Figure 7.1.9)

Triangulation corresponding to M = N = 5

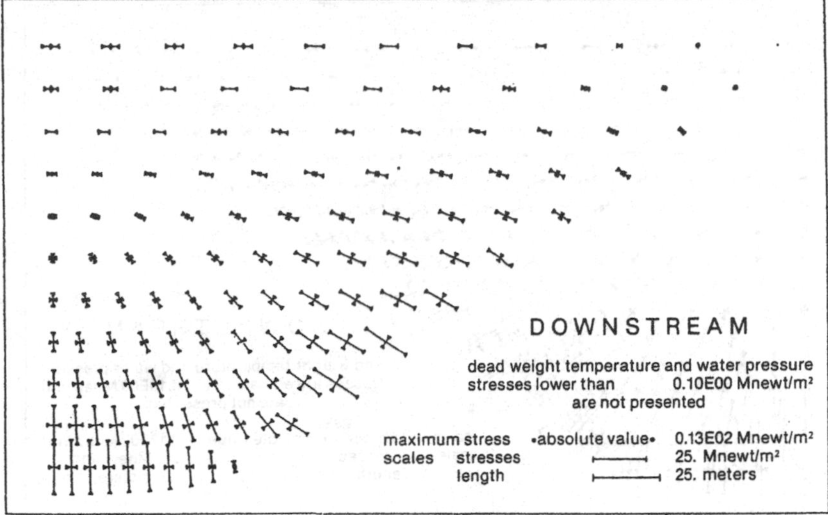

Figure 7.2.6 : Stress distribution on downstream face (50 triangles)
(compare with Figure 7.1.8)

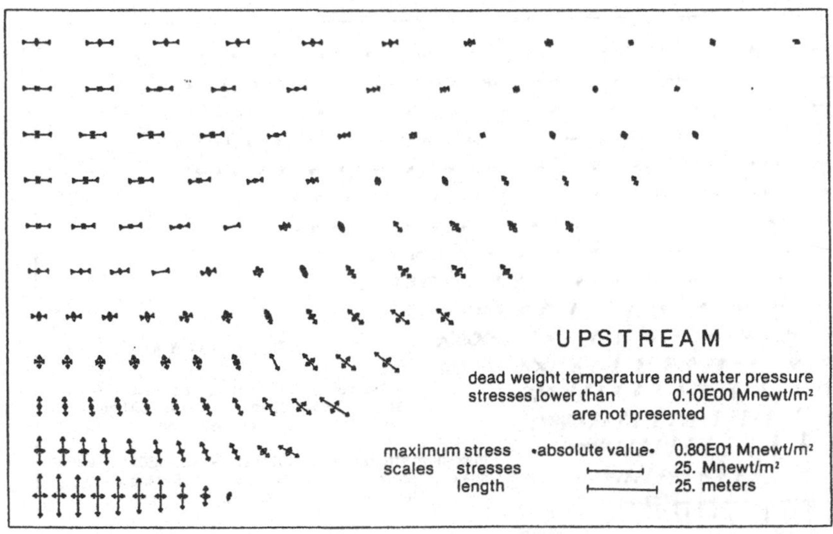

Figure 7.2.7 : Stress distribution on upstream face (50 triangles)
(compare with Figure 7.1.9)

Triangulation corresponding to M = N = 8

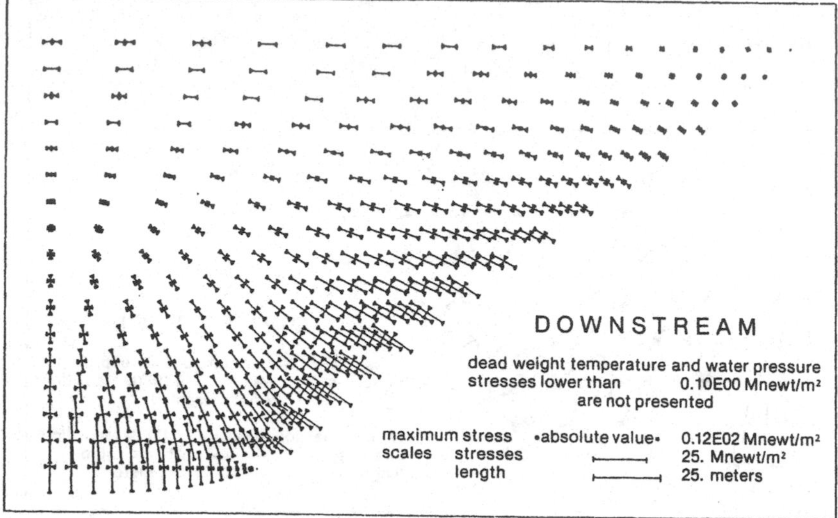

Figure 7.2.8 : Stress distribution on downstream face (128 triangles)
(compare with Figure 7.1.8)

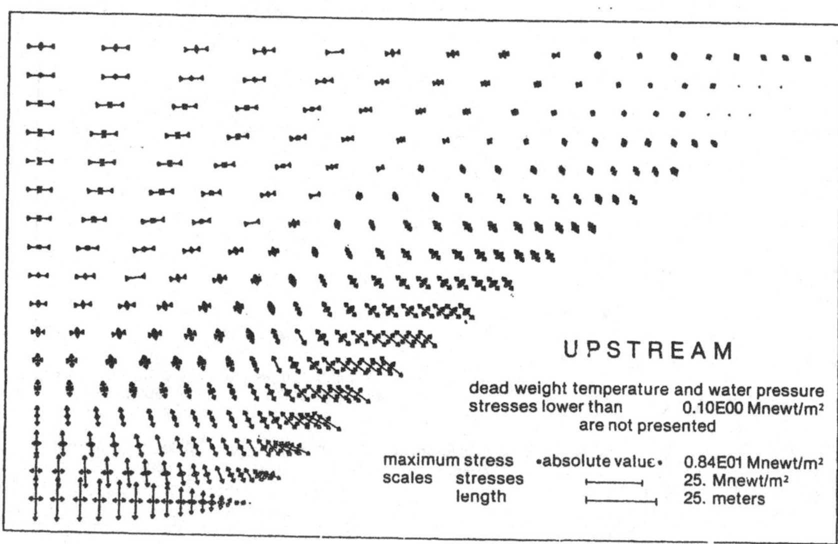

Figure 7.2.9 : Stress distribution on upstream face (128 triangles)
(compare with Figure 7.1.9)

Irregular mesh (see Figure 7.2.1)

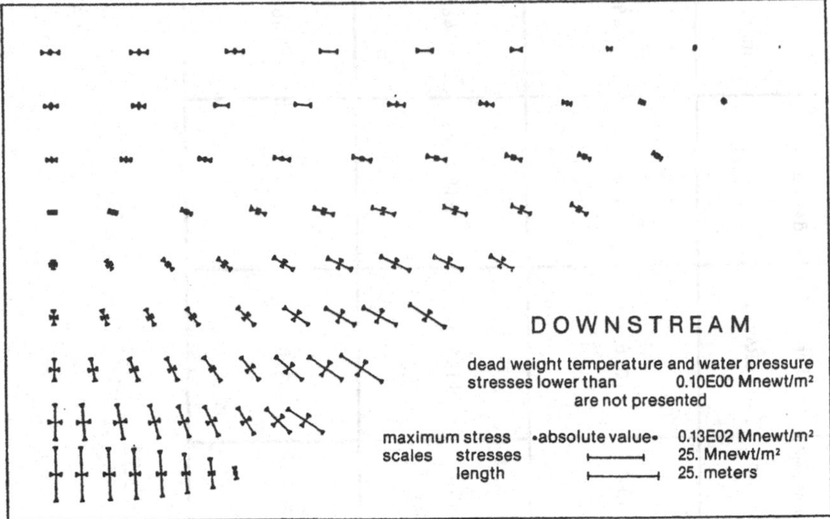

Figure 7.2.10 : Stress distribution on downstream face (32 triangles)
(compare with Figure 7.1.8)

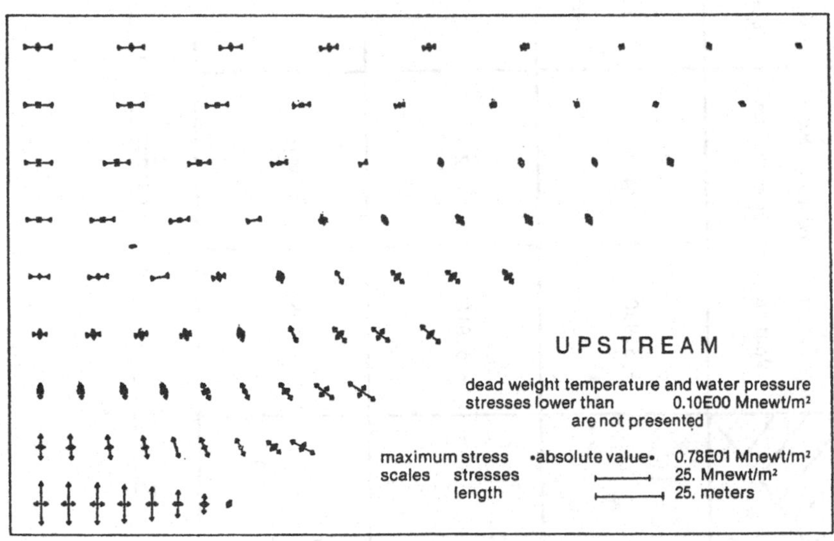

Figure 7.2.11 : Stress distribution on upstream face (32 triangles)
(compare with Figure 7.1.9)

Scheme in use	Top of crown cantilever- *Horizontal Stresses* (MN/m^2)		Base of crown cantilever- *Horizontal Stresses* (MN/m^2)		Computing time (on CRAY 1)	Nb. of degrees of freedom	Nb. of unknowns
	Upstream	Downstream	Upstream	Downstream			
12 points exact for P_6	- 5.930	- 4.228	1.662	- 2.450	10.6 s	618	463
16 points exact for P_8	- 5.916	- 4.216	1.662	- 2.450	11.4 s	618	463
25 points exact for P_{10}	- 5.905	- 4.210	1.661	- 2.450	13.8 s	618	463

Table 7.3.1 : Effect of changes of numerical integration scheme

BIBLIOGRAPHY

ADAMS, R.A. [1975] : *Sobolev Spaces*, Academic Press, New York.

AHMAD, S. ; IRONS, B.M. ; ZIENKIEWICZ, O.C. [1970] : Analysis of thick and thin shell structures by curved finite elements, *Internat. J. Numer. Methods Engrg.* 2, pp. 419-451.

ARGYRIS, J.H. ; FRIED, I. ; SCHARPF, D.W. [1968] : The TUBA family of plate elements for the matrix displacement method, *Aero. J. Royal Aeronautical Society* 72, pp. 701-709.

ARGYRIS, J.H. ; HAASE, M. ; MALEJANNAKIS, G.A. [1973] : Natural geometry of surfaces with specific reference to the matrix displacement analysis of shells, I, II and III, *Proc. Kon. Ned. Akad. Wetensch., Series* B, 76, pp. 361-410.

ARGYRIS, J.H. ; LOCHNER, N. [1972] : On the application of the SHEBA shell element, Comput. Methods Appl. Mech. Engrg. 1, pp. 317-347.

ASHWELL, D.G. ; GALLAGHER, R.H. [1976] : *Finite elements for thin shells and curved members*, J. Wiley and Sons, London.

BEGIS, D. ; PERRONNET, A. [1980] : Présentation du Club MODULEF, *Notice 50*, Version 3.2, INRIA.

BELL, K. [1969] : A refined triangular plate bending element, Internat. J. Numer. Methods Engrg. 1, pp. 101-122.

BERNADOU, M. [1978] : *Sur l'analyse numérique du modèle linéaire de coques minces de W.T. KOITER, Thèse d'Etat,* Université Pierre et Marie CURIE, Paris.

BERNADOU, M. [1980] : Convergence of conforming finite element methods for general shell problems, *Internat. J. Engrg. Sci.*, 18, pp 249-276.

BERNADOU, M. ; BOISSERIE, J.M. [1978a] : Implémentation de l'élément fini d'ARGYRIS - Exemples, *Rapport IRIA-LABORIA* 301.

BERNADOU, M. ; BOISSERIE, J.M. [1978b] : Sur l'implémentation de problèmes généraux de coques, *Rapport IRIA-LABORIA* 317.

BERNADOU, M. ; BOISSERIE, J.M. ; HASSAN, K. [1980], Sur l'implémentation des éléments finis de HSIEH-CLOUGH-TOCHER, complet et réduit, *Rapports de Recherche INRIA,* 4.

BERNADOU, M. ; CIARLET, P.G. [1976] : Sur l'ellipticité du modèle linéaire de coques de W.T. Koiter, in *Computing Methods in Applied Sciences and Engineering* (R. Glowinski and J.L. Lions, Editors), pp. 89-136, Lecture Notes in Economics and Mathematical Systems, Vol. 134, Springer-Verlag, Berlin.

BERNADOU, M. ; DUCATEL, Y. [1978] : Méthodes d'éléments finis avec intégration numérique pour des problèmes elliptiques du quatrième ordre, *Rev. Française Automat. Informat. Recherche Opérationnelle, Analyse Numérique*, 12, Numéro 1, pp. 3-26.

BERNADOU, M. ; HASSAN, K. [1981] : Basis functions for general HSIEH-CLOUGH-TOCHER triangles, complete or reduced, *Internat. J. Numer. Methods Engrg.*, 17, pp. 784-789.

BOGNER, F.K. ; FOX, R.L. ; SCHMIT, L.A. [1965] : The generation of interelement compatible stiffness and mass matrices by the use of interpolation formulas, in *Proceedings of the Conference on Matrix Methods in Structural Mechanics*, Wright Patterson A.F.B., Ohio.

BRAMBLE, J.H. ; HILBERT, S.R. [1970] : Estimation of linear functionals on Sobolev spaces with application to Fourier transforms and spline interpolation, *SIAM J. Numer. Anal.* 7, pp. 113-124.

CARLSON, D.E. [1972] : Linear Thermoelasticity, *Handbuch der Physik*, Vol. VI a-2, Springer-Verlag, Berlin, pp. 297-345.

CIARLET, P.G. [1976] : Conforming finite element methods for the shell problems, in *The Mathematics of Finite Elements and Applications II* (J.R. Whiteman, Editor), Academic Press, London, pp. 105-123.

CIARLET, P.G. [1978] : *The Finite Element Method for Elliptic Problems*, North-Holland, Amsterdam.

CIARLET, P.G. ; DESTUYNDER, P. [1979] : "Approximation of three-dimensional models by two-dimensional models in plate theory", *Energy Methods in Finite Element Analysis*, Edited by R. Glowinski, E.Y. Rodin, O.C. Zienkiewicz ; John Wiley & Sons, Chichester, pp. 33-45.

CLOUGH, R.W. ; JOHNSON, C.P. [1970] : Finite element analysis of arbitrary thin shells, *Proceedings ACI Symposium on concrete thin shells*, New-York, pp. 333-363.

CLOUGH, R.W. ; TOCHER, J.L. [1965] : Finite element stiffness matrices for analysis of plates in bending, in *Proceedings of the Conference on Matrix Methods in Structural Mechanics*, Wright Patterson A.F.B. Ohio.

COSSERAT, E. and F. [1909] : *Théorie des corps déformables*, Herman, Paris.

COUTRIS, N. [1976] : Théorème d'existence et d'unicité pour un problème de flexion élastique de coques dans le cadre de la modèlisation de P.M. Naghdi. *C.R. Acad. Sci. Paris, Sér. A*, 283, pp. 951-953.

COUTRIS, N. [1978] : Théorème d'existence et d'unicité pour un problème de coque élastique dans le cas d'un modèle linéaire de P.M. Naghdi, *Rev. Française Automat. Informat. Recherche Opérationnelle, Analyse Numérique*, 12, n° 1, pp. 51-58.

COWPER, G.R. [1973] : Gaussian quadrature formulas for triangles, *Internat. J. Numer. Methods Engng.*, 7, n° 3, pp. 405-408.

COYNE & BELLIER, [1977] : Barrage de GRAND'MAISON, *Dossier préliminaire.*

CRAINE, R.E. [1968] : Spherically symmetric problem in finite thermo-elastostatics, *Quart. J. Mech. Appl. Math.*, XXI, Pt. 3, pp. 279-291.

DESTUYNDER, P. [1980] : *Sur une justification mathématique des théories de plaques et de coques en élasticité linéaire*, Thèse d'Etat, Université Pierre et Marie Curie, Paris.

DESTUYNDER, P. ; LUTOBORSKI, A. [1980] : A penalty method for the BUDIANSKY-SANDERS shell model. *Rapport Interne*, 67, Centre de Mathématiques Appliquées de l'Ecole Polytechnique.

DUPUIS, G. [1971] : Application of Ritz method to thin elastic shell analysis, *J. Appl. Mech.*, 71-APM-32, pp. 1-9.

DUPUIS, G. ; GOEL, J.-J. [1970a] : Finite elements with a high degree of regularity, *Internat. J. Numer. Methods Engrg.* 2, pp. 563-577.

DUPUIS, G. ; GOEL, J.-J. [1970b] : A curved finite element for thin elastic shells, *Internat. J. Solids and Structures* 6, pp. 1413-1428.

DUVAUT, G. ; LIONS, J.L. [1972] : *Les Inéquations en Mécanique et en Physique*, Dunod, Paris.

ERICKSEN, J.L. [1960] : Appendix- Tensor Fields, Handbuch der Physik, Vol. III/1, Springer-Verlag, Berlin, pp. 793-858.

FRAEIJS DE VEUBEKE, B. [1965a] : Bending and stretching of plates, in *Proceedings of the Conference on Matrix Methods in Structural Mechanics*, Wright Patterson A.F.B., Ohio.

GERMAIN, P. [1973] : *Cours de Mécanique des Milieux Continus*, Tome 1, Théorie Générale, Masson, Paris.

GREEN, A.E. ; ADKINS, J.E. [1970] : *Large Elastic Deformations*, second edition, revised by A.E. GREEN, Clarendon Press, Oxford.

GREEN, A.E. ; NAGHDI, P.M. [1970] : Non-isothermal theory of rods, plates and shells, *Internat. J. Solids and Structures*, 6, pp. 209-244.

GREEN, A.E. ; NAGHDI, P.M. [1978] : On thermal effects in the theory of shells, *Office of Naval Research Report* n° UCB/AM-78-4.

HAASE, M. [1977] : On the construction of Gaussian quadrature formulae for triangular regions, *ISD Internal Report*, Univ. Stuttgart.

HAMMER, P.C. ; STROUD, A.H. [1956] : Numerical integration over simplexes, *Math. Tables Aids Comput.*, 10, pp. 137-139.

HILLION, P. [1977] : Numerical integration on a triangle, *Internat. J. Numer. Methods Engrg.*, 11, pp. 797-815.

HLAVÁČEK, I. ; NEČAS, J. [1970] : On inequalities of Korn's type I and II, *Arch. Rational Mech. Anal.*, 36, pp. 305-334.

JENNINGS, A. [1971] : Solution of variable bandwidth positive definite simultaneaous equations, *Comput. J.*, 14, p. 446.

JOHNSON, C. [1975] : "On finite element methods for curved shells using flat elements", *Numerische Behandlung von Differentialgleichungen*, International Series of Numerical Mathematics, Vol. 27, Birkhaüser Verlag, Basel and Stuttgart, pp. 147-154.

KIRCHHOFF, G. [1876] : *Vorlesungen über Mathematische Physik*, Mechanik, Leipzig.

KNOWLES, N.C. ; RAZZAQUE, A. ; SPOONER, J.B. [1976] : "Experience of finite element analysis of shell structures", *Finite Elements for thin shells and curved members*, Edited by D.G. ASHWELL, R.H. GALLAGHER, J. Wiley & Sons, London, pp. 245-262.

KOITER, W.T. [1966] : On the nonlinear theory of thin elastic shells, *Proc. Kon. Ned. Akad. Wetensch.*, B 69, pp. 1-54.

KOITER, W.T. [1970] : On the foundations of the linear theory of thin elastic shells, *Proc. Kon. Ned. Akad. Wetensch.*, B 73, pp. 169-195.

KOITER, W.T. ; SIMMONDS, J.C. [1973] : Foundations of shell theory, *Proceedings of Thirteenth International Congress of Theoretical and Applied Mechanics*, Moscow, Août 1972, Springer-Verlag, Berlin, pp. 150-176.

LANDAU, L. ; LIFCHITZ, E. [1967] : *Théorie de l'Elasticité*, Editions MIR, Moscou.

LAURSEN, M.E. ; GELLERT, M. [1978] : Some criteria for numerically integrated matrices and quadrature formulas for triangles, *Internat. J. Numer. Methods Engng.*, 12, pp. 67-76.

LIONS, J.L. ; MAGENES, E. [1968] : *Problèmes aux limites non homogènes et applications*, Vol. 1, Dunod, Paris.

LOVE, A.E.H. [1934] : *The Mathematical Theory of Elasticity*, Cambridge University Press.

LYNESS, J.N. ; JESPERSEN, D. [1975] : Moderate degree symmetric quadrature rules for the triangle, *J. Inst. Math. Appl.*, 15, pp. 19-32.

MALVERN, L.E. [1969] : *Introduction to the mechanics of a continuous medium*, Prentice-Hall, Inc., Englewood Cliffs.

Mc CONNELL, A.J. [1931] : *Applications of the absolute differential calculus*, Blackie, London and Glasgow.

NAGHDI, P.M. [1963] : Foundations of elastic shell theory, *Progress in Solid Mechanics*, Vol. 4, North-Holland, Amsterdam, pp. 1-90.

NAGHDI, P.M. [1972] : The Theory of Shell and Plates, *Handbuch der Physik*, Vol. VI a-2, Springer-Verlag, Berlin, pp. 425-640.

NAYLOR, D.J. ; STAGG, K.G. ; ZIENKIEWICZ, O.C. [1975] : *Criteria and assumptions for numerical analysis of dams*, Proceedings of an International symposium held at Swansea, U.K., 8-11 September, 1975, University College, Swansea.

NEČAS, J. [1967] : *Les Méthodes Directes en Théorie des Equations Elliptiques*, Masson, Paris.

ODEN, J.T. ; REDDY, J.N. [1976] : *An Introduction to the Mathematical Theory of Finite Elements*, Wiley Interscience, New-York.

PERRONNET, A. [1979], The Club MODULEF : "A library of subroutines for finite element analysis", *Computing Methods in Applied Sciences and Engineering*, Edited by R. Glowinski and J.L. Lions, Lectures Notes in Mathematics, 704, Springer-Verlag, Berlin, pp. 127-153.

RUTTEN, H.S. [1973] : *Theory and design of shells on the basis of asymptotic analysis*, Rutten + Kruisman, Consulting engineers, Rijswijk, Holland.

RYDZEWSKI, J.R. [1965] : *Theory of Arch Dams*, Pergamon Press, Oxford.

SANDER, G. [1969] : Applications de la méthode des éléments finis à la flexion des plaques, *Collection des publications de la Faculté des Sciences de Liège*, 15.

STEPHAN, E. ; WEISSGERBER, V. [1978] : Zur approximation von schalem mit hybriden elementen, *Computing*, 20, n° 1, pp. 75-95.

STROUD, A.H. [1971] : *Approximate Calculation of Multiple Integrals*, Prentice-Hall, Englewood Cliffs.

TRUESDELL, C. [1953] : The physical components of vectors and tensors, *Z. Angew. Math. Mech.*, 33, n° 10-11, pp. 345-356.

WEMPNER, G.R. ; ODEN, J.T. ; KROSS, D. [1968] : Finite Element Analysis of Thin Shells, *J. Engrg. Mech. Div.*, ASCE, Vol. 94, n° EM6, pp. 1273-1294.

ZIENKIEWICZ, O.C. ; TAYLOR, R.L. ; TOO, J.M. [1971] : Reduced integration technique in general analysis of plates and shells, *Internat. J. Numer. Methods Engrg.*, 3, pp. 275-290.

ZLÁMAL, M. [1974] : Curved elements in the finite element method, Part II, *SIAM J. Numer. Anal.*, 11, pp. 347-362.

GLOSSARY OF SYMBOLS

Symbol	Name or description	Place of definition or first occurence
a	$= \det (a_{\alpha\beta})$	(1.1.18)(1.5.13)
$a(.,.)$	bilinear form associated with the shell strain energy	(1.3.26)
$a_h(.,.)$	approximate bilinear form	(2.2.4)(2.4.15)
a_i	vertices of a triangle	pages 30, 66
$\vec{a}_\alpha$	covariant basis of the tangent plane to the undeformed middle surface	(1.1.2)
$\vec{a}^\alpha$	contravariant basis of the tangent plane to the undeformed middle surface	(1.1.6)
$\vec{\bar{a}}_\alpha$	covariant basis of the tangent plane to the deformed middle surface	(1.3.3) (1.3.15) (1.3.18)
$a_{\alpha\beta}$	first fundamental form	(1.1.4)(1.5.12)
$a^{\alpha\beta}$	elements of the inverse matrix $[a_{\lambda_\Gamma}]^{-1}$	(1.1.7)(1.5.14)
$\vec{a_3} = \vec{a}^3$	normal vector to the undeformed middle surface	(1.1.3)

Symbol	Name or description	Place of definition or first occurence
$\vec{\bar{a}}_3 = \vec{\bar{a}}^3$	normal vector to the deformed middle surface	(1.3.4)(1.3.19)
A_A	matrix of shape functions for ARGYRIS triangle	(3.1.19), Figure 3.1.2
A_{C_i}	matrix of shape functions for complete H.C.T. triangle	(3.1.33), Figure 3.1.3
A_{R_i}	matrix of shape functions for reduced-H.C.T. triangle	(3.1.41), Figure 3.1.4
AS	denotes parameters which are antisymmetric with respect to the plane x = 0	(6.4.17) to (6.4.20)
A_{IJ}	matrix coefficient associated with the bilinear form	(1.5.3)(1.5.10)
b_i	midpoints of the sides of a triangle	page 30
$b_{\ell,K}$	nodes of a numerical integration scheme	(2.2.2)(2.4.10)
$b_{\alpha\beta}$	second fundamental form	(1.1.5)(1.5.15)
$b_{\alpha}^{\cdot\beta}=b^{\beta}_{\cdot\alpha}=b^{\beta}_{\alpha}$	mixed components of the second fundamental form	(1.1.8)
$b^{\alpha\beta}$	contravariant components of the second fundamental form	(1.1.8)
$\bar{b}^{\beta}_{\alpha}$	mixed components of the second fundamental form of the *deformed* middle surface	(1.3.5)(1.3.7)

Symbol	Name or description	Place of definition or first occurence
B_i	matrix associated with the linear form $f_h(.)$	(3.3.9)(3.4.9) (3.5.9)(3.6.9)
c_i	the point of intersection of a line from a vertex a_i perpendicular to the opposite side of a triangle	page 30
c^{α}_{β}	mixed components of the third fundamental form	(1.5.17)
$\mathcal{C}$	configuration of the undeformed shell	(1.2.2) ; page 10
$\mathcal{C}^*$	configuration of the deformed shell	page 10
$\mathcal{C}^m, m=0,1$	space of functions m times continuously differentiable	page 29
ds	line element	(4.3.16)
$d\mho$	area element	(5.2.8)
dS	area element of the middle surface	(1.1.19)
∂S	boundary of the middle surface	page 5
∂S_o	clamped part of the boundary of the middle surface	(1.3.28)
dV	volume element	(5.3.18)
d_i	components of a change matrix	(3.1.23)
d_4	component of a change matrix	(3.1.24)

Symbol	Name or description	Place of definition or first occurence
DG	set of global degrees of freedom	(3.2.6)
D_A	change matrix associated with ARGYRIS triangle	(3.1.22)(3.2.7)
D_{C_i}	change matrix associated with complete-H.C.T. triangle	(3.1.36)
D_{R_i}	change matrix associated with reduced-H.C.T. triangle	(3.1.44)
DA_{C_i}	change matrices	(3.3.4)(3.4.4)
DA_{R_i}	change matrices	(3.5.4)(3.6.4)
DG_i	sets of global degrees of freedom	(3.3.2)
$DLGL_A$	set of *global* degrees of freedom associated with ARGYRIS triangle	(3.1.20)
$DLGL_{C_i}$	set of *global* degrees of freedom associated with complete-H.C.T. triangle	(3.1.34)
$DLGL_{R_i}$	set of *global* degrees of freedom associated with reduced-H.C.T. triangle	(3.1.43)
$DLLC_A$	set of *local* degrees of freedom associated with ARGYRIS triangle	(3.1.18)
$DLLC_{C_i}$	set of *local* degrees of freedom associated with complete-H.C.T. triangle	(3.1.32)

Symbol	Name or description	Place of definition or first occurence
$DLLC_{R_i}$	set of *local* degrees of freedom associated with reduced-H.C.T. triangle	(3.1.40)
DT	change matrix	(3.2.8)
DT_i	change matrices	(3.3.3)
$D^{\alpha}v(a)$	α-th (FRECHET) derivative of a function v at a point a	page 3
e	thickness of the shell	(1.2.1)(4.4.3)
$\vec{e}_i$, i=1,2,3	orthonormal basis	page 5
$e^{\alpha\beta}$, $e_{\alpha\beta}$	permutation matrices	(1.1.17)
E	YOUNG's modulus	page 15, (5.3.2) page 123
E(.)	error functionals associated with the use of a numerical integration scheme	(2.4.12)
$\hat{E}$(.)		(2.4.11)
E_d	strain energy of the arch dam	(5.3.4)
E_{d1}	pure deformation component of the strain energy	(5.3.6)
E_{d2}	thermal component of the strain energy	(5.3.7)(5.3.20)
$\mathcal{E}^2$	Euclidean plane	pages 3, 5
$\mathcal{E}^3$	Euclidean space	page 5

Symbol	Name or description	Place of definition or first occurence
$f(.)$	linear form giving the work of external loads	(1.3.29)(5.4.2)
$f_h(.)$	approximate linear form	(2.2.5)(2.4.16)
F	matrix associated with the linear form f(.)	(1.5.20)(5.5.2) (6.4.21)
F_K	mapping which associates triangle K with the reference triangle $\hat{K}$	(2.4.1)
g	$= \det (g_{ij})$	(5.1.2)
$g_\circ$	gravitational acceleration	(5.1.2)(5.2.2) page 123
$\vec{g}_i$	covariant basis of the undeformed shell	(1.2.3)
$\vec{\bar{g}}_i$	covariant basis of the deformed shell	(1.3.5)
g_{ij}	metric tensor of the undeformed shell	(1.3.7)
$\bar{g}_{ij}$	metric tensor of the deformed shell	(1.3.7)
g_{IJ}	functions which give A_{IJ}	(1.5.3)
$\vec{G}$	volume density of the weight of the dam	(5.1.1)
h	$= \max_{K \in \mathcal{C}_h} h_K$	(2.1.2)

Symbol	Name or description	Place of definition or first occurence
h_K	= diam (K)	(2.1.1)
$H^m(\Omega)$	SOBOLEV space	page 4
$\mathfrak{J}$	upstream wall of the dam	page 113
$J(\vec{v})$	energy of the shell associated with a displacement field $\vec{v}$	(1.4.6)(5.4.1)
k	thermal conductivity coefficient	page 116
(K,P_K,Σ_K)	finite element	page 40
$(\hat{K},\hat{P},\hat{\Sigma})$	reference finite element	page 40
ℓ_i	length of the side $a_{i-1}\ a_{i+1}$	(3.1.16)
LAMBD	matrix of barycentric coordinates and their derivatives	(3.2.9)
LAMBD_i	matrix of barycentric coordinates and their derivatives	(3.3.5)(3.4.5) (3.5.5)(3.6.5)
M	position of a particle of the undeformed shell	page 11
M	number of subdivisions of the interval [0,1] of the x-axis	page 124
$\bar{M}$	position of the particle M after deformation	page 11
M_i	matrices associated with the approximate bilinear form	(3.3.7)(3.4.7) (3.5.7)(3.6.7)

Symbol	Name or description	Place of definition or first occurence
M^{α}_{β}	matrix associated with the change of curvature tensor	(1.5.9)
$\vec{n}$	unit normal vector to the upstream wall directed to the external part of the arch dam	page 112 (5.2.4)
$\vec{n}$	unit normal vector to the boundary Γ_{o}	(6.3.4)
n_i	parameters used to construct matrix D_A	(3.1.22)(3.1.25)
N	number of subdivisions of the interval [0,1] of the y-axis	page 124
p	pressure upon the upstream wall of the dam	(5.2.2)
p_A	basis polynomials for ARGYRIS triangle	(3.1.19)
P	position of a particle of the undeformed middle surface	page 11
$\bar{P}$	position of the particle P after deformation	page 11
$P_m(K)$	space of all polynomials of degree $\leq m$	pages 30, 40
P_K	space of shape functions for element K	page 30
P_S	symmetry plane of the arch dam	page 132

Symbol	Name or description	Place of definition or first occurence
R_C	curvature radius of the middle surface of the arch dam	(4.3.17)
R_{C_i}	shape functions for the complete-H.C.T. triangle	(3.1.33)
R_{R_i}	shape functions for the reduced-H.C.T. triangle	(3.1.41)
s	arc length of the middle line	(4.4.1)
S	denotes parameters which are symmetric with respect to the plane $x = 0$	(6.4.17) to (6.4.20)
S	upper-triangular matrix	page 143
$S, \bar{S}$	middle surface of the shell	page 5
$\vec{t}$	unit tangent vector to the boundary Γ_0	(6.3.3)
$\overrightarrow{t_1}, \overrightarrow{t_2}$	half-tangents at a salient point	(6.3.7)
T	change of temperature field	(5.3.10)(5.3.12)
$\mathscr{C}_h$	triangulation of the polygonal domain Ω	(2.1.1)
T_{do}	downstream wall temperature	(5.3.11)(7.1.1)
T_{up}	upstream wall temperature	(5.3.11)(7.1.1)
T_G	work of gravitational loading of the arch dam	(5.1.3)(5.1.6) (5.1.7)

Symbol	Name or description	Place of definition or first occurence
T_H	work of water pressure loading	(5.2.6)(5.2.10)
T_1	mean-value of the upstream and downstream wall temperatures	(5.3.13)
T_2	moment of order 1 associated with the upstream and downstream wall temperatures	(5.3.14)
$T_{\alpha\|\gamma}$	covariant derivatives	(1.1.11)
$T_{\alpha\beta\|\gamma}$	covariant derivatives	(1.1.12)
$\vec{u}$	displacement field of the particles of the middle surface S	(1.3.13)
U	column matrix of components of the displacement $\vec{u}$ and their derivatives	(1.5.1)(1.5.2)
$\vec{U}$	displacement field of the particles of $\mathcal{C}$	page 11 ; (1.3.12)(1.3.20)
$\vec{U}_h$	approximate displacement field	(6.6.2)
V	column matrix of components of the displacement $\vec{v}$ and their derivatives	(1.5.1)(1.5.2)
$\vec{V}$	space of admissible displacements	(1.4.2)(5.4.3)
$\vec{\tilde{V}}$	antisymmetric subspace of $\vec{V}$	(6.4.10)
$\vec{\approx{V}}$	symmetric subspace of $\vec{V}$	(6.4.11)
$\vec{V}_h$	approximate space of admissible displacements	page 29 ; (2.1.6)(2.1.7)

Symbol	Name or description	Place of definition or first occurence
V_{h1}	subspace of X_{h1}	(2.1.4)
V_{h2}	subspace of X_{h2}	(2.1.5)
$W^{m,p}(\Omega)$	SOBOLEV spaces	page 3
X_{h1}	finite element subspace of the space $H^1(\Omega)$	(2.1.3)
X_{h2}	finite element subspace of the space $H^2(\Omega)$	(2.1.3)
Z	coordinate along the vertical	Fig. 4.1.3 ; (4.2.3)
$Z_\circ$	height of the arch dam	(4.2.5)
Z_1	level of water in the reservoir	page 151
α	parameter of the arch dam	(4.2.5)
$\bar{\alpha}$	coefficient of linear expansion	(5.3.3) ; page 124
γ^*_{ij}	strain tensor of the shell $\mathcal{C}$	(1.3.6) ; page 115
$\gamma^{*i}_{\ j}$	mixed components of the strain tensor of the shell	(6.7.3)
$\gamma_{\alpha\beta}$	strain tensor of the middle surface	(1.3.10)(1.3.21)
γ^α_β	mixed components of the strain tensor of the middle surface	(1.3.25)(6.7.1)
Γ	boundary of the domain Ω	page 5

Symbol	Name or description	Place of definition or first occurence
$\Gamma_\circ$	$\Gamma_\circ$ = clamped part of the boundary Γ	page 16
$\Gamma^{\alpha}_{\beta\gamma}$	CHRISTOFFEL's symbols	(1.1.10)(1.5.16) (4.3.24)
$\Gamma_{\alpha\beta\gamma}$	CHRISTOFFEL's symbols	(4.3.18) to (4.3.23)
$\delta_{\alpha\beta}$	KRONECKER's symbols	(1.5.7)
Δ	$\|\Delta\|$ = 2.(area of the triangle)	(3.1.5)
$\Delta_{\alpha\beta}$	$\Delta_{11} = \Delta_{22} = 0$; $\Delta_{12} = \Delta_{21} = 1$	(1.5.7)
$\vec{\varepsilon}_\alpha$	orthonormal basis in the Euclidean plane $\mathbb{R}^2$	(3.1.2)
$\varepsilon_{\alpha\beta}, \varepsilon^{\alpha\beta}$	ε-system for the middle surface	(1.1.16)
η_i	eccentricity parameters	(3.1.14)
θ	parameter used in the definition of the middle surface of the arch dam	(4.2.3)
$\theta_\circ$	parameter used in the definition of the middle surface of the arch dam	(4.2.3)(4.2.5)
θ	temperature function of the arch dam	(5.3.1)
$\theta_\circ$	initial uniform temperature of the arch dam	(5.3.1)

Symbol	Name or description	Place of definition or first occurence
λ_i	barycentric coordinates of the triangle	(3.1.3) to (3.1.5)
Λ^{α}_{β}	matrix associated with the strain tensor of the middle surface	(1.5.6)
ν	POISSON's coefficient	(1.3.24)(5.3.2) page 123
(ξ^1,ξ^2)	system of orthonormal coordinates of the $\mathcal{E}^2$-plane	page 3
(ξ^1,ξ^2,ξ^3)	system of curvilinear coordinates for space $\mathcal{E}^3$	page 9
$(\xi^1,\xi^2,\bar{\xi}^3)$	new system of curvilinear coordinates used in the description of the deformed shell	page 12
Π_K	interpolation operator	page 30
Π_{KA}	interpolation operator associated with ARGYRIS triangle	(3.1.17)(3.1.28)
Π_{K_iC}	interpolation operator associated with complete-H.C.T. triangle	(3.1.31)(3.1.37)
Π_{K_iR}	interpolation operator associated with reduced-H.C.T. triangle	(3.1.39)(3.1.45)
$\overrightarrow{\Pi_h v}$	$\vec{V}_h$-interpolant of a function $\vec{v} \in \vec{V}$	(2.2.6)
$\rho_\circ$	parameter used in the definition of the middle surface of the arch dam	(4.2.6)

Symbol	Name or description	Place of definition or first occurence
ρ_1	mass density of the concrete in the undeformed configuration	(5.1.2) ; page 123
ρ_2	mass density of the water	(5.2.2) ; page 123
$\bar{\rho}_{\alpha\beta}$	change of curvature tensor of the middle surface	(1.3.11)(1.3.22)
$\bar{\rho}^{\alpha}_{\beta}$	mixed components of the change of curvature tensor of the middle surface	(1.3.25)(6.7.2)
σ^{*ij}	stress tensor of the shell	(5.3.1)(6.7.4)
Σ_K	set of degrees of freedom	page 30
$\vec{\phi}$	mapping used in the definition of the middle surface	(1.1.1)
Φ	functional used in the (infinitesimal) rigid body motion lemma	(1.6.3)
ψ	equivalent norm on the space $(H^1(\Omega))^2 \times H^2(\Omega)$	(1.6.1)(1.6.2)
ω	angle used in order to take into account boundary conditions	(6.3.3)(6.3.4)
$\omega_{\ell,K}$	weights of the numerical integration scheme	(2.2.2)(2.4.10)
Ω	open bounded subset in a plane $\mathcal{E}^2$ which is used as a reference domain	pages 1, 5 ; Figure 4.2.2

Symbol	Name or description	Place of definition or first occurence
$\tilde{\Omega}$	reference domain of the middle surface of the arch dam before simplifications	Figure 4.2.2
Ω_1	half-domain Ω	Figure 6.4.1
$\Vert\cdot\Vert_{m,p,\Omega}$	norm on the space $W^{m,p}(\Omega)$, that is $\Vert v\Vert_{m,p} = \left(\sum_{\vert\alpha\vert\le m} \int_\Omega \vert D^\alpha v\vert^p \, d\xi \right)^{1/p}$, $1\le p<\infty$	page 3
$\vert\cdot\vert_{m,p}$	semi-norm on the space $W^{m,p}(\Omega)$, that is $\vert v\vert_{m,p} = \left(\sum_{\vert\alpha\vert = m} \int_\Omega \vert D^\alpha v\vert^p \, d\xi \right)^{1/p}$, $1\le p<\infty$	page 3
$\Vert\cdot\Vert^*_{m,p,\Omega}$	norm in the dual of the space $W^{m,p}(\Omega)$	Lemma 2.4.1
$\Vert\cdot\Vert_{m,\Omega}$	$= \Vert\cdot\Vert_{m,2,\Omega}$	page 4
$\vert\cdot\vert_{m,\Omega}$	$= \vert\cdot\vert_{m,2,\Omega}$	page 4
$((\cdot,\cdot))_{m,\Omega}$	scalar product on the space $H^m(\Omega) = W^{m,2}(\Omega)$, that is $((u,v))_{m,\Omega} = \sum_{\vert\alpha\vert\le m} \int_\Omega D^\alpha u \, D^\alpha v \, d\xi$	page 4
$\vert\cdot\vert$	norm on the space $L^2(\Omega)$, that is $\vert v\vert = \vert v\vert_{0,\Omega}$	(1.6.2)
	<u>or</u>	
	Euclidean norm in $\mathcal{E}^3$	(1.1.3)

Symbol	Name or description	Place of definition or first occurence
	usual scalar product in $\mathcal{E}^3$, that is $(\vec{a},\vec{b}) = \vec{a}\cdot\vec{b}$	page 6
$\hookrightarrow$	inclusion with continuous injection	page 4
$\|$	covariant derivatives ; for instance $T_{\alpha\|\gamma}$; $T_{\alpha\beta\|\gamma}$	(1.1.11)(1.1.12)
(*)	this exponent indicates a difference between tensors in $\mathcal{E}^3$ and tensors defined on the middle surface	(1.3.6)
[0]	corresponding degree of freedom is known	page 128
[1]	corresponding degree of freedom is unknown	page 128

INDEX